Total Parenteral Nutrition in the Hospital and at Home

Editor

Khursheed N. Jeejeebhoy, M.B., Ph.D.
Professor
Department of Medicine
University of Toronto
Chief
Division of Gastroenterology
Toronto General Hospital
Toronto, Ontario, Canada

With the Assistance of
Alan Bruce-Robertson, M.D.
Toronto General Hospital
Toronto, Ontario, Canada

CRC Press
Taylor & Francis Group
Boca Raton London New York

CRC Press is an imprint of the
Taylor & Francis Group, an informa business

First published 1983 by CRC Press
Taylor & Francis Group
6000 Broken Sound Parkway NW, Suite 300
Boca Raton, FL 33487-2742

Reissued 2018 by CRC Press

Library of Congress Cataloging in Publication Data

Total parenteral nutrition in the hospital and at home.

Bibliography: p.
Includes index.
1. Parenteral feeding. 2. Hospital care. 3. Home nursing. I. Jeejeebhoy, Khursheed N. II Bruce-Robertson, Alan. [DNLM: 1. Parenteral feeding—Methods. 2. Parenteral feeding—Nursing. WB 410 J44t]
RM224.T673 1983 616.8'55 82-9397
ISBN 0-8493-6120-6 AACR2

A Library of Congress record exists under LC control number: 82009397

Publisher's Note
The publisher has gone to great lengths to ensure the quality of this reprint but points out that some imperfections in the original copies may be apparent.

Disclaimer
The publisher has made every effort to trace copyright holders and welcomes correspondence from those they have been unable to contact.

ISBN 13: 978-1-315-89823-0 (hbk)
ISBN 13: 978-1-351-07733-0 (ebk)

Visit the Taylor & Francis Web site at http://www.taylorandfrancis.com and the
CRC Press Web site at http://www.crcpress.com

INTRODUCTION

The recognition by the medical profession of the importance of nutrition as an integral part of patient management is a natural evolution in the progress of therapeutics. Earlier efforts in the direction of better cardiorespiratory support, attention to fluid and electrolyte status, and infection control have resulted in increased survival of seriously sick patients who nevertheless remain sufficiently incapacitated so as to be unable to eat a normal diet, a disability which together with the catabolic effects of their illness results in malnutrition. Furthermore, in patients with gastrointestinal disease the propensity for developing malnutrition is especially enhanced because of their inability to eat and absorb oral nutrients. Consequently if this state of affairs continues then recovery is hindered even if the underlying disease is cured.

It is therefore obvious that good patient management necessitates the use of an alternative route of nutritional support in patients unable to eat or absorb an oral diet. This alternative is parenteral nutrition, which is the subject of this book. While there are many texts on the subject of parenteral nutrition, very few if any are directed to the practical details of organizing the delivery of parenteral nutrition from a multidisciplinary point of view. In this publication we present the practice of parenteral nutrition as viewed by a team of a physician, nurse, and pharmacist.

THE EDITOR

Khursheed N. Jeejeebhoy, M.B., Ph.D., was born in Burma. He did his undergraduate medical training at the Christian Medical College, Vellore, India. He then completed postgraduate training at the West Middlesex Hospital and the Royal Post-Graduate Medical School, London. Dr. Jeejeebhoy was appointed Head of the Medical Division at the Radiation Medicine Centre, Bombay, where he was involved with research into sprue and nutrition. He was appointed to the Staff of the Department of Medicine at the University of Toronto in 1968, where he is presently Head of the Division of Gastroenterology at the Toronto General Hospital, Chairman of the Nutrition Review Committee, and has been the Coordinator of the Gastroenterology Training Program of the University of Toronto.

His main research interest is in parenteral nutrition and role of nutrition in patients with gastrointestinal diseases. He is the author of 132 papers and 47 book chapters.

CONTRIBUTORS

Zane Cohen, M.D.
Assistant Professor
Department of Surgery
University of Toronto
Toronto, Ontario, Canada

Khursheed N. Jeejeebhoy, Ph.D.
Chief, Division of Gastroenterology
Toronto General Hospital
Toronto, Ontario, Canada

Gail Kennedy, R.N., B.Sc.N.
Kimberley, British Columbia, Canada

Bernard Langer, M.D., FRCS(C)
Head, Division of General Surgery
Toronto General Hospital
Professor and Chairman
Department of Surgery
University of Toronto
Toronto, Ontario, Canada

George Tsallas, B.Sc.
Supervisor of Manufacturing
Pharmacy Department
Toronto General Hospital
Toronto, Ontario, Canada

TABLE OF CONTENTS

Chapter 1

ORGANIZATION OF THE NUTRITIONAL SUPPORT TEAM

Khursheed N. Jeejeebhoy

TABLE OF CONTENTS*

* Reading List follows Chapter 4.

I. INTRODUCTION

During the past few years it has become evident that the diagnosis and treatment of malnutrition and its complications requires the cooperative effort of different health professions in addition to the traditional role of the physician. This synthesis can only be effective if the different health professions work in close harmony, with clearly defined roles and a mutual understanding of their respective interests. To achieve these ends various centers concerned with clinical nutrition have recognized that it is necessary to formally constitute a defined team of the relevant health professions.

To appreciate the benefits of such a team vs. a situation where nutritional support is given by independent groups, it is necessary to examine the various types of expertise that are required to deliver the best care to the patient. The physician has the overall knowledge of the clinical condition, the degree of malnutrition, the need for nutritional support, and the most effective route for providing such support.

When the physician has decided to use parenteral nutrition it is necessary to have sterile solutions compounded to meet the needs of an individual patient. Alternatively, where a basic mixture is applicable to the treatment of a variety of patients, the formula can be standardized and bulk-manufactured by special aseptic techniques. In any case these special services and pharmaceutical techniques among many other considerations make close and frequent consultations with pharmacy an important part of the delivery of nutritional support to the patient.

The ultimate delivery of this therapy depends upon sound nursing practices, carried out by a team of nurses who understand the principles of artificial, parenteral and enteral, nutrition. The total management of the patient depends on their understanding the therapy given in relation to nutritional requirements, careful attention to safe procedures for the delivery of parenteral nutrition and monitoring for possible complications. To fulfill these aims continuing education and liaison with nurses and within the nursing teams is of utmost importance.

Finally the physician, the nurse, pharmacist, and dietician must interact with each other to provide harmonized care. The only way this can be done is to form a team which can work together to achieve designated goals. Since enteral nutrition is beyond the scope of this text the team concept will be primarily directed to the delivery of parenteral nutrition.

A. Members of the Team

The core team should consist of a physician, nurse, and pharmacist. This team will interface with an extended group as indicated below.

B. Organization of the Team

There are two different ways of organizing the TPN team: decentralized and centralized.

1. Decentralized Team

In this system there is a core team as described above which establishes standards of nutritional assessment, protocols for the management of nutritional problems, and guidelines for formulating the nutritional prescription as well as standardizing pharmaceutical formulations and establishing standards of nursing and medical care for the delivery of parenteral nutrition. The core team will also educate other hospital staff responsible for patient care.

The core team operates by providing a consultative service to aid the primary physician and ward nurses, and the clinical pharmacist to formulate and deliver TPN in a safe and effective way to the patient.

The main advantage of this system is that patients remain under the care of their primary physician and TPN is thus given in the context of total care. In addition there is continuity of the care under the same nursing, ward, and house staff. Furthermore by working through communication rather than exclusive control of patients there is less tension and an enhanced education about the nutritional needs of patients. The disadvantage of this method is the fact that quality control can only be exercised if the staff looking after the patient faithfully and willingly cooperates with the team. Also, a lot more effort is required by the team members to make the decentralized approach work.

2. Centralized Team

With a centralized team, patients requiring parenteral nutrition are taken over by the core team members and their assistants. To meet this objective patients may be transferred to a special geographic area under the care of the team physician or they may be located in any ward where the specifics related to parenteral nutrition would be the sole responsibility of the team and not the ward staff.

The chief advantage of this system is the easy enforcement of high standards of care in regard to the delivery of TPN. In a small hospital with limited expertise this approach is economical in terms of resources.

The disadvantages include a loss of the perspective that TPN is only a part of the total care of the patient which may produce decisions conflicting with those best made by a group looking after the patient as a whole. This applies particularly to large hospitals looking after patients with complex problems. Furthermore, the rest of the hospital staff remains ignorant about nutritional needs being an important part of patient care.

II. GOALS AND OBJECTIVES OF THE TEAM

1. Develop guidelines and standards of practice for the institution.
2. Develop a protocol which defines the role of the physician, nurse, and pharmacist and enforces safe practices to be used by the nursing units and house staff.
3. Provide a consultation service for nutritional problems.
4. Undertake regular education by means of lectures, seminars, discussions of other personnel looking after the patient, and attendance at nursing and pharmacy staff rounds.
5. Conduct research studies and update techniques in the light of new knowledge.
6. Evaluate the results of the nutritional service and take corrective measures when problems are recognized.
7. Develop services to undertake a home TPN program in those institutions where the volume and type of problem justifies this modality.

III. THE TEAM AND DELIVERY OF TPN

The delivery of TPN by a team involves two main considerations. First, the team must create a protocol and prepare themselves and the relevant nursing and pharmacy services to be able to accept and care for patients on TPN. Hence, as a background, the nurse and pharmacist from the core team will develop protocols, discuss and obtain approval of these protocols by the team, and then implement them by providing in-service education and organization. The nature of this process is given in detail in the sections of the book concerned with the role of the nurse and the pharmacist. Secondly, when this preparation is complete, then the practice goes into effect as given below.

The primary physician initiates the nutritional support process by asking for a consultation from the team physician. The team physician, after appropriate nutritional and clinical assessment, then formulates a plan of treatment which is discussed with the primary physician and the ward nursing staff. The nutrition prescription is sent to the pharmacy where it is checked for compatibility and appropriateness of the formulation. Simultaneously, the team nurse provides in-service education to the ward nursing staff. Then the patient receives the infusions. After the treatment is started the patient is monitored for response to therapy and for the development of complications. The TPN nurse reviews the procedures of the nursing care to ensure that protocol is strictly observed in regard to catheter care and infusion techniques. The pharmacist ensures that sterile solutions with appropriate additives reach the floor, maintains a close eye on the orders for compatibilities and dosage, and reviews the progress of the patient through the cooperative efforts of the team.

Another approach is for the team to implement orders to start TPN by the primary physician, without reference to the physician running the TPN service. With this system the request is automatically complied with as long as standard practices are followed. The physician running the TPN team only comes into the picture when unusual solutions or techniques are requested. This approach has many disadvantages, the most important being the inability to educate medical personnel on advances in nutritional therapy in an ongoing way. Clearly the physician running the TPN service is more likely to be familiar with recent advances in a way not possible for other nonexpert physicians. Hence this system will retard the translation of recent advances into practice for patient care.

Chapter 2

INDICATIONS

Khursheed N. Jeejeebhoy

TABLE OF CONTENTS*

* Reading List follows Chapter 4.

I. INTRODUCTION

The indications for parenteral nutrition should be approached from two points of view. First, does the patient need nutritional support and second, is the parenteral route the preferred way of giving nutrients to the patient? The need for nutritional support will now be considered in detail.

II. MALNUTRITION OR POTENTIAL FOR MALNUTRITION

It is obvious that if the patient is malnourished then an effort has to be made to give additional nutrients. On the other hand, it is not so obvious that patients need active intervention to avoid malnutrition if they are suffering from a disease that prevents them from nourishing themselves. Hence malnutrition or the potential development of malnutrition should be indications for nutritional support.

III. INJURY AND SEPSIS

It was shown by Cuthbertson and colleagues 50 years ago that injury resulted in enhanced nitrogen loss. In addition, it is a common clinical observation that patients with injury and sepsis will develop marked wasting and hypoalbuminemia. Hence, the occurrence of injury and sepsis should alert the physician that nutritional support may be required.

IV. BOWEL REST

In a number of gastrointestinal (GI) diseases keeping the patient NPO (nil per os) results in an improvement of symptoms. This effect is most obvious in patients with Crohn's disease, where this form of therapy is conceptually equivalent to diverting the luminal stream from the inflamed bowel. However, keeping the patient NPO will clearly result in malnutrition unless the patient is given parenteral nutrition. The combination of the NPO state and parenteral nutrition is referred to as bowel rest therapy. This therapy has been employed in treating Crohn's disease, pancreatitis, bowel fistulas, and resistant sprue-like syndromes. While controlled trials to prove the efficacy of bowel rest in all these conditions is lacking, nevertheless there is strong evidence from prospective studies that in Crohn's disease there is a favorable effect of such therapy in reducing both the activity of the disease and the size of inflammatory masses. The results are less striking when bowel rest is used to close bowel fistulas in patients with Crohn's disease, at least on a longer-term basis.

V. CANCER THERAPY

Cancer is associated with weight loss and in a number of studies it has been shown that weight loss is associated with a poor prognosis. Hence, it is not surprising that attempts were made to see if nourishment would improve the outlook. Initial claims of parenteral nutrition being exceptionally valuable in cancer treatment have not stood up to critical analysis. Controlled trials, where available, have not shown any improvement in survival in patients given parenteral nutrition nor is the effect of chemotherapy improved by parenteral nutrition. Recently Mullen et al. have shown that in selected patients with low albumin and cancer, parenteral nutrition may reduce postoperative morbidity. The role of nutritional support in cancer cachexia awaits further trials.

VI. ADJUNCTIVE THERAPY PREOPERATIVELY

There are numerous clinical reasons to believe that operation on the malnourished patient is associated with increased morbidity. The few controlled studies done in this area have supported this concept. Hence there is good reason to give nutritional support to malnourished patients who may be undergoing surgery. Since the use of low-calorie protein-free solutions has been shown to result in a negative nitrogen balance, it has been suggested that all patients be given nutritional support pre- and postoperatively; a concept which will increase the complexity of managing the otherwise well-nourished patients having elective procedures. This concept of universal parenteral nutrition has no support at the present time and is not recommended.

VII. GROWTH RETARDATION

Growth retardation is a frequent and unfortunate consequence of Crohn's disease and results in cessation of growth and maturation in adolescents with this condition. Several studies have clearly indicated that this process is entirely due to the poor intake of calories because of anorexia and discomfort following meals. Despite the presence of growth retardation, the bowel disease in these patients may not be clinically active. This apparent quiescence of the disease is the result of poor intake reducing or preventing the occurrence of obstruction and diarrhea. The use of nutritional support, both oral and parenteral, has resulted in catch-up growth and onset of puberty.

VIII. ORAL OR PARENTERAL NUTRITION

Having considered whether the patient does need nutritional support the next question is whether it is to be given orally or parenterally. This decision depends upon a number of factors which will now be considered. It is axiomatic that patients who can nourish themselves orally should be nourished by that route. However, this simplistic approach may not be easily applied to an individual patient and may lead to undesirable results. For example even though the GI tract may be superficially intact, there may be functional derangements such as altered motility resulting in anorexia, early satiety, distension after meals, gastric retention and vomiting — all being the result of systemic illness. When such patients are encouraged to eat increased amounts to meet their enhanced metabolic needs they may be unable to comply and may actually become even more nauseated and develop an aversion to food. Under these circumstances it has been suggested that oral feeding can be undertaken by infusing defined formula diets through a fine-bore tube. While in theory this technique looks promising, in practice this approach does not permit the infusion of all necessary calories and proteins on the very first day of feeding; the patient has to adapt to the diet by initially receiving smaller volumes of a more dilute diet which is then gradually built up to the appropriate volume of a full-strength diet. Without such adaptation nausea, vomiting, and diarrhea may occur. In the process of adaptation between 3 to 7 days may be spent on a suboptimal diet, which would be undesirable in very sick patients with serious malnutrition. From the foregoing it is obvious that patients should be selected for parenteral nutrition both on the absence of a functional GI tract and on the basis of serious malnutrition with intolerance to an oral diet where the urgency of the problem may necessitate rapid refeeding despite food intolerance and GI motility disorders.

IX. HOME PARENTERAL NUTRITION

An important step in the development of parenteral nutrition was its application to patients who had suffered irreversible bowel injury of a magnitude which prevented them from deriving nourishment from an oral diet. This clinical problem involved not only the use of parenteral nutrition to keep them alive but, in addition, the development of methods to allow it to be used at home in such a way that they could carry on with a normal social life and be gainfully employed. This process has been called home parenteral nutrition or HPN. The indications for this therapy are

1. Total jejuno-ileal resection, where oral nutrition is impossible.
2. Massive small bowel resection, resulting in the patient being unable to maintain oral nutrition despite careful attention to the use of defined formula diets and other supplements. In this context it is important to realize that in some patients an oral diet may be possible by the use of extreme measures, such as eating every 2 hr and defecating about as frequently. Clearly, such a patient is a social cripple and is a candidate for HPN despite being able to get by with oral feeds. In other cases HPN may be used to get patients home and working while waiting for the remaining bowel to adapt after a massive resection. When adaptation occurs, HPN is stopped.
3. Chronic small bowel obstruction not amenable to surgical alleviation, where eating causes pain and vomiting thus limiting oral alimentation. This may occur in Crohn's disease with multiple strictures not amenable to a simple surgical solution. It may also occur with scleroderma and the pseudo-obstruction syndrome. In fact, in the latter condition HPN has revolutionized the outlook.
4. End-jejunostomy syndrome. This is a condition in which the patient has lost those areas of the small and large bowel where intestinal contents are concentrated. Hence, there is copious loss of isotonic bowel contents after eating. The patient absorbs nutrients, but suffers severe fluid and electrolyte deficiencies, especially magnesium deficiency. In such patients the provision of parenteral fluid and electrolyte supplements has once again been a significant advance in therapy.
5. Growth retardation. In some patients with Crohn's disease and growth retardation, in whom oral feeding even with defined formula diets is not possible or not tolerated, a regimen of HPN will be beneficial and permit catch-up growth.

Chapter 3

NUTRIENT NEEDS

Khursheed N. Jeejeebhoy

TABLE OF CONTENTS*

* Reading List follows Chapter 4.

I. INTRODUCTION

The human body is built around a matrix of protein which is not stored to any significant extent as a surplus commodity, most of it being present as an essential constituent of the body. Body protein is not a static entity but is constantly being turned over by the processes of synthesis and catabolism. During these processes there are obligatory losses of protein nitrogen as urea, the amount depending upon both the protein intake and the metabolic rate. Starvation causes a fall in insulin levels, resulting in the mobilization of fatty acids from adipose tissue stores to provide fuel for energy requirements. This process aids in preserving protein from being broken down for this purpose. Unfortunately, despite the availability of energy from fat stores during starvation, there is continuing protein catabolism which will gradually result in the wasting of essential tissues (Figure 1). Hence starvation is associated with a progressive loss of body protein — a nonexpendable commodity — and injury only adds to the rate at which such loss occurs. This fact is apparent from Table 1. This table shows the total body nitrogen measured directly by neutron activation in patients with various degrees of clinical malnutrition. From the table it is obvious that body nitrogen falls with increasing malnutrition. Since this loss is of vital body structures, the first priority of nutritional support is to restore the lost protein — directly by providing amino acids and indirectly by giving carbohydrate and fat to meet energy needs, thus preventing the amino acids from being catabolized for energy. In addition, the provision of electrolytes, trace elements, and vitamins aid protein synthesis in ways which will be indicated later. Furthermore it is obvious that apart from supplying nutrients for the restoration of structure, it is also necessary to provide nutrients for meeting energy needs, fluid and electrolyte homeostasis, and other metabolic processes.

II. PROTEIN METABOLISM

A. Factors Affecting Protein Utilization

Protein utilization has been estimated by observing the amount of protein and energy needed to create a positive nitrogen balance. Furthermore, recent papers have emphasized the role of nitrogen balance to determine nutrient needs during parenteral nutritional therapy. Hence it is appropriate to analyze the value and interpretation of nitrogen balance data. The human adult has finite cellularity of muscle and viscera which in turn determines the body nitrogen content; hence a well-nourished adult cannot go into positive nitrogen balance unless there is a tissue increase which is neither a part of muscle nor of viscera. Such an increase has been shown to occur mainly with deposition of adipose tissue and is believed to be the nitrogen required to permit the development of support structures such as connective tissue, blood vessels, and the increase in muscle fiber size necessary to support the additional weight due to adiposity. Experimentally this concept has been supported by the studies of Keys et al. in volunteers and by Elwyn in patients on parenteral nutrition. Munro had previously shown that nitrogen excretion in urine is dependent upon the previous nitrogen intake. Hence the amount of nitrogen required to maintain balance in any individual depends on his/ her previous intake of dietary nitrogen. When the individual on a high protein diet starves or reduces protein intake, the previously high urinary nitrogen loss continues for a few days despite reduced intake, resulting in negative nitrogen balance, the magnitude of which is directly proportional to the previous level of intake. The higher the nitrogen intake, the greater the negative balance upon starvation, and vice versa. However, after a few days the nitrogen excretion falls until it is equal to the new lower level of protein intake or equivalent to at least 2 mg/cal of metabolic rate. Therefore, protein intake necessary for nitrogen balance will rise with an increase in metabolic

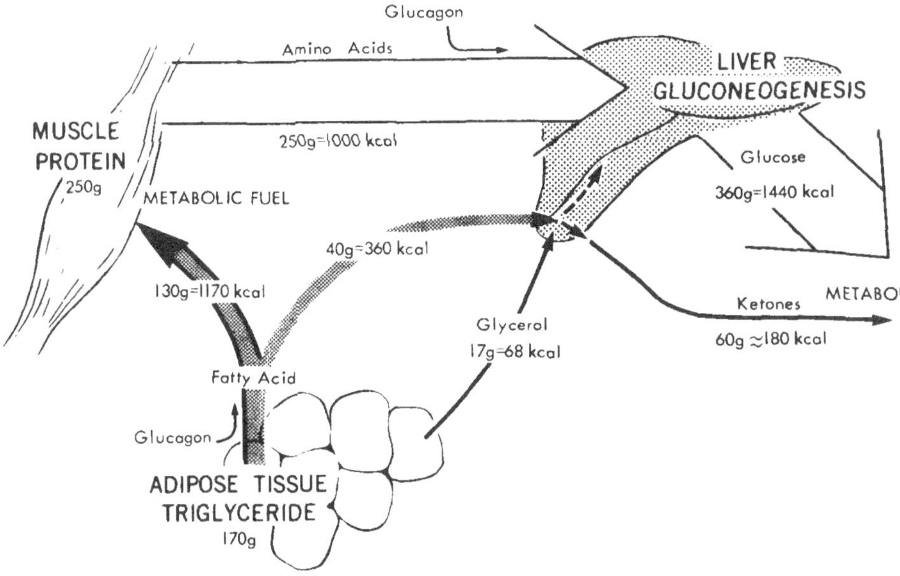

FIGURE 1. Metabolic fuel fluxes (2600 kcal) showing route by which protein catabolism continues.

rate. However the cumulative initial negative nitrogen balance referred to above has a limit of about 50 to 60 g (out of a total of about 1800 to 2000 g of total body nitrogen) after which excretion falls to the low value of about 2 mg/cal of metabolic rate. Thus, to meet continuing losses it is necessary to replace about 2 mg/cal of metabolic rate, and more is needed to rebuild lost tissue in the malnourished. Recognizing the above principles can aid in reconciling various studies with apparently contradictory results. Calloway and Spector, and more recently Richardson et al., showed that in healthy adults it is necessary to give at least 1 g protein and 45 kcal/kg body weight to maintain nitrogen balance. In contrast, in malnourished and postoperative subjects, apart from the need for energy substrate (glucose or fat) Greenberg et al. noticed that the quantity of nitrogen given was an important determinant of nitrogen balance; provided that 2 g/kg amino acids were given per day to such patients, a positive nitrogen balance was seen even when total calories were less than the metabolic requirements. Similar results have been obtained in postoperative patients by Bozetti. Hence by giving amino acids alone it is possible to induce nitrogen balance in depleted adults. By contrast the well-nourished adult needs excess energy to deposit fat before a positive nitrogen balance is attained. Experimental proof for this contention can be obtained from the studies of Yeung et al. who showed that patients receiving total parenteral nutrition gained a lot of weight as compared with those on a more modest calorie intake with an elemental diet. However, all this weight gain was shown to be water or fat and nitrogen retention was equivalent. Even in patients with burns who are traditionally regarded as having high caloric needs, Burke et al. have shown that when the protein intake was as high as 2.78 g/kg/day, the effect of adding glucose calories on protein synthesis or catabolism was minimal (Table 2). In other studies by Collins et al. the total body nitrogen after major surgery was maintained as well with an amino acid infusion as with amino acid and glucose.

B. Protein Requirements

From the foregoing discussion of protein metabolism it is clear that the requirement is dependent on a number of metabolic factors such as the previous nutritional status, degree of nutritional depletion, provision of nonprotein energy, and the rate of desired

Table 1
EFFECT OF MALNUTRITION ON TOTAL BODY POTASSIUM (TBK) AND TOTAL BODY NITROGEN (TBN)

Clinical assessment of degree of malnourishment	Mean TBK (g)		Mean TBN (kg)	
	Male	Female	Male	Female
None	129	74	2.0	1.5
Moderate	119	76	1.7	1.2
Severe	91	61	1.6	1.1

Table 2
PROTEIN SYNTHESIS AND BREAKDOWN AT DIFFERENT RATES OF GLUCOSE INFUSION IN A PATIENT WITH 65% BURN SURFACE AREA

Rate of i.v. glucose infusion mg/kg/min	Protein (3m/kg/day)	
	Synthesis	Breakdown
2.5	5.2	3.5
5.0	5.0	3.0
9.0	5.0	3.0

Modified from Burke, J. F., Wolfe, R.R., Mullany, C.J., Mathews, D.W., and Bier, D.M., *Ann. Surg.*, 190, 274, 1979.

repletion. In general, the more protein given the less the effect of nonprotein energy intake and the greater the degree of nitrogen retention in the depleted patient. In addition to these purely metabolic factors the question is often asked whether intravenously given amino acids are as effective in promoting nitrogen retention as oral proteins, and if all amino acid preparations are comparable in regard to their ability to promote nitrogen retention. To answer this question Patel et al. did a study of the effect of casein hydrolysate, given intravenously in graded amounts, on the nitrogen balance and compared the results with those obtained from oral casein and oral hydrolysate therapy. This study demonstrated that the amino acid composition of the hydrolysate given intravenously was not optimal and resulted in a lower balance when compared with that obtained with casein given orally. However the balance was comparable with that where the same hydrolysate was administered orally. The finding that the nitrogen balance during the intravenous administration of casein hydrolysate was the same as the one observed when it was given orally, keeping all other parameters constant, indicated that oral and intravenous amino acids were handled identically as far as gross nitrogen economy was concerned. It was surmised that the difference between the hydrolysate and whole casein in that study was due to a loss of sulfur-containing and aromatic amino acids during the process of hydrolysis. To prove this point Anderson et al. devised and infused an amino acid mixture enriched in these two groups of amino acids. Using a similar study design they showed that with a suitable

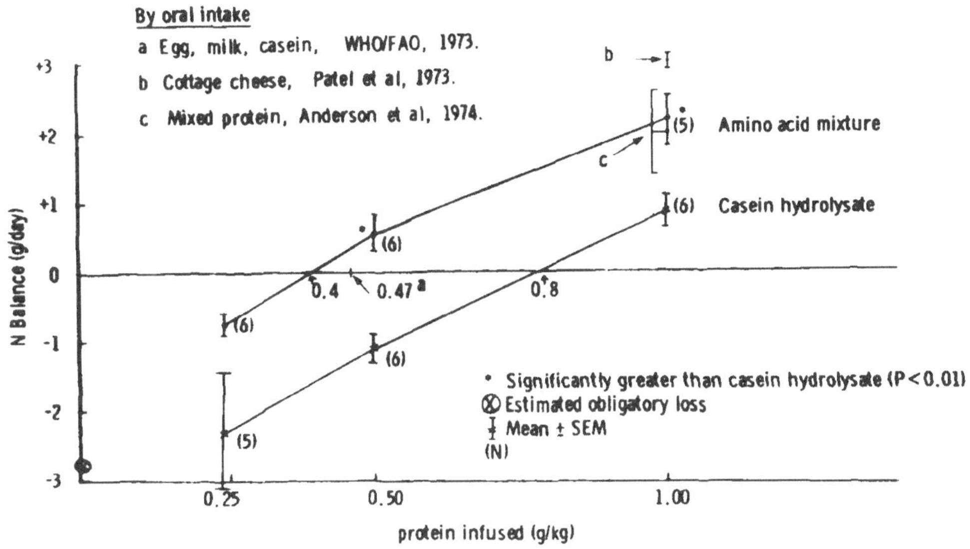

FIGURE 2. Nitrogen balance in patients receiving different amounts of a given amino acid mixture intravenously.

amino acid composition the intravenously administered amino acids were as effective in promoting a positive nitrogen balance as oral protein, and about 1 g/kg ideal body weight was sufficient to promote a positive nitrogen balance and an increase in total body nitrogen in malnourished patients (Figure 2).

C. Recommendations for Intravenous Amino Acids

The nitrogen requirement for balance in stable adults is only 0.4 g/kg/day and this was also found to be true for intravenous amino acids in the study by Anderson et al. However, in malnourished patients it is not only desirable to attain balance but also to gain nitrogen. This gain was linear over a range of 0.25 to 2 g/kg/day in several studies. Furthermore an increase in metabolic rate increases nitrogen loss, raising the basal needs for balance. Finally, the higher the nitrogen intake the less dependent is balance on energy intake in depleted patients. On the other hand, poor renal and hepatic function will reduce tolerance to amino acid loads. Taking all these factors into account it is desirable to give as a first approximation 1 to 1.5 g/kg of ideal body weight per day of a balanced amino acid mixture and to monitor the response of plasma proteins and urea nitrogen excretion.

III. ENERGY METABOLISM

A. Energy Requirements

Energy requirements are dependent on a number of factors which include the body surface area derived from height and weight, age and sex. It can be predicted with reasonable accuracy by the Harris-Benedict equation for normal man. In a study done by Shike on 21 adults of various heights and weights the calculated data agreed with the observed energy consumption calculated by indirect calorimetry with a variation of ±5%. The equations are given below.

Men: kcal/24 hr = 66.473 + 13.7516 × wt (kg) + 5.0033 × ht (cm) − 6.775 × age (yr)

Women: kcal/24 hr = 655.0955 + 9.5634 × wt (kg 9 + 1.8496 × ht (cm) − 4.6756 × age (yr)

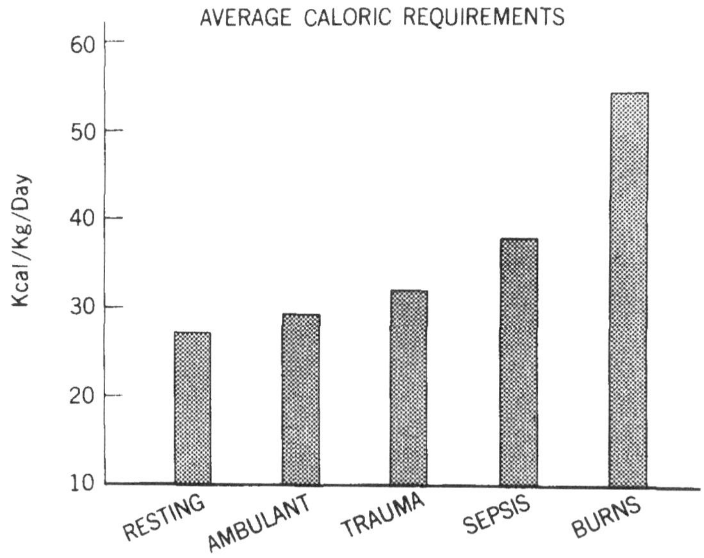

FIGURE 3. Resting energy requirements measured in patients suffering injury, sepsis, or burns.

These considerations apply to the basal energy expenditure (BEE) and to these values must be added the specific dynamic action of food to give the resting energy expenditure (REE) in the bedridden patient. Elwyn has stated that REE can be approximated from the BEE by increasing it by 10%. In contrast, malnutrition reduces the BEE to an extent which exceeds that expected simply on the basis of weight loss by an additional 35%. Injury, sepsis, and especially burns increase energy requirements by approximately 30, 60, and 100%, respectively (Figure 3). However these are peak values taken at the height of injury and in practice these do not hold for the average increase over the period of illness in injured-septic patients. For example, the mean increase in the metabolic rate of injured septic patients studied by Askanazi et al. exceeded the expected value by only 14%. Similarly Roulet et al. found an increase of only 12.9% in critically ill patients in respiratory failure. In any case it should be recognized that the rise in metabolic rate of even 60%, when referred to the BEE which is about 25 kcal/kg/day, works out to 40 kcal/kg/day or 2800 kcal in the 70 kg man. Hence there is little evidence for the need to use 4000 to 6000 kcal in the vast majority of injured-septic patients with the exception of those with major burns.

B. Source of Energy Substrate — Carbohydrate and Fat

Gamble in his classical studies showed that feeding glucose reduced the excretion of nitrogen in the urine in fasting subjects, demonstrating the protein-sparing effect of glucose. Subsequently, in the injured-septic patient Kinney et al. showed that there was an increase in energy consumption and that nitrogen loss paralleled the metabolic rate. Based on the known protein-sparing effects of glucose and the observed increased metabolic rate and nitrogen loss in such subjects, parenteral nutrition using large amounts of glucose was advocated for treating injured-septic patients with the belief that this therapy would promote nitrogen retention or reduce catabolism, especially in such patients. This concept received support from the studies of Woolfson and Allison and Long et al., who showed that in burned patients glucose was the only substrate capable of reducing nitrogen excretion and improving nitrogen balance. However, in the injured-septic patient, Kinney and colleagues had shown that the respiratory quotient was low and indicated that in such patients fat was the principle source of energy. This finding was at variance with the above concepts.

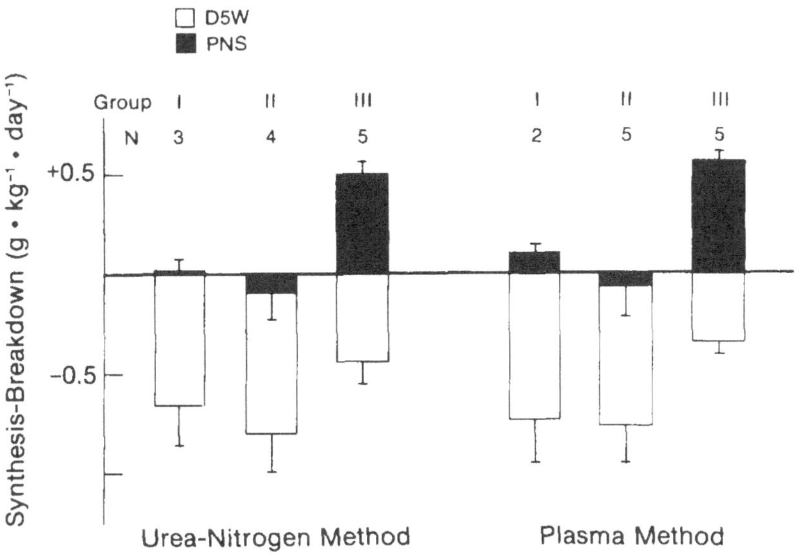

FIGURE 4. Net protein synthesis in patients receiving parenteral nutritional support (PNS) in the form of (I) amino acids, (II) amino acids plus glucose, or (III) amino acids and equicaloric amounts of glucose and lipid; and comparing this synthesis with that when receiving only 5% dextrose in water (D5W). This last and PNS I supplied 4 kcal/kg, and PNS II and III 40 kcal/kg ideal body weight. (Modified from Roulet, M. et al., *J. Parenteral Enteral Nutr.*, 4 (6), 582, 1980.)

Recently the paramount importance of glucose has come into question, especially for the majority of patients with GI disease and intra-abdominal sepsis who need TPN. Jeejeebhoy et al. showed that, in contrast to the observation of Long et al., the patients with GI disease utilize glucose and fat equally well and both promote nitrogen retention to a similar extent. This finding was confirmed by several other investigators. In a controlled trial, Yeung and colleagues showed that glucose alone does not increase total body nitrogen above that seen with protein alone, and the weight gain noted with such a source of nonprotein calories was largely due to water retention. Later the same group, in a controlled trial, showed that a glucose-lipid mixture promoted nitrogen retention and, by contrast, glucose alone did not do so. Parallel observations by Wolfe et al., Burke et al., and Askanazi et al. showed that the injured-septic patient, including those with burns, had an obligatory need for fat energy and such patients continued to oxidize significant amounts of fat despite providing all nonprotein calories as glucose. Furthermore they did not utilize all the infused glucose calories for energy. The excess glucose calories are converted to fat in the liver, and result in the hepatic steatosis seen by Jeejeebhoy et al. and Messing et al. when glucose is used as an exclusive source of nonprotein calories but not when a glucose-lipid mixture is used. Finally Askanazi et al. showed that infusing hypercaloric glucose in the injured-septic patient resulted in hypermetabolism, a rise in catecholamine excretion, and increased oxygen consumption and CO_2 production. The last effect was especially of concern in patients with respiratory failure. Finally, Roulet and Jeejeebhoy (Figure 4) have shown in a controlled trial that acutely sick patients in respiratory failure showed only a significant increase in net protein synthesis over catabolism as measured by ^{14}C-leucine turnover studies, when a mixed substrate of fat and carbohydrate is infused. In contrast, infusing equivalent amounts of amino acids without calories, or with glucose calories alone, only maintained balance between synthesis and catabolism — observations in accordance with those of Hill and colleagues. Hence it is now clear that the injured septic patient needs a mixed substrate of glucose and fat for meeting energy needs.

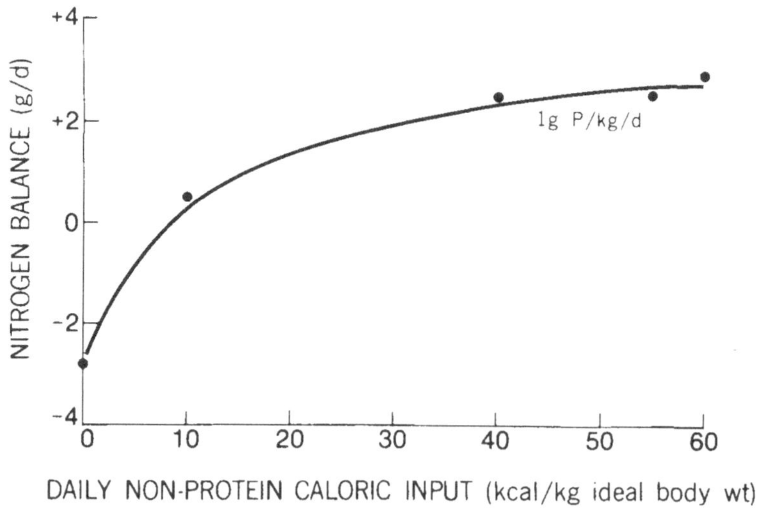

FIGURE 5. Nitrogen balance in patients (ill, but not immediately postoperative) receiving daily and intravenously the equivalent of 1.0 g of protein/kg ideal body weight (IBW)/day, along with glucose-lipid mixtures supplying nonprotein calories at levels ranging from 0 to 60 kcal/kg IBW/d. (From Greene, H. L., Holliday, M. A., and Munro, H. N., Eds., *Clinical Nutrition Update: Amino Acids*, American Medical Association, Chicago, 1977, 71. With permission.)

C. Recommendations for Energy Intake

The next question is the magnitude of the optimal calorie intake in the majority of patients. Hitherto, based on the concept that these patients are grossly hypercatabolic, 4000 to 5000 kcal were often given to sick patients. In studies using a graded input of calories, Jeejeebhoy and associates found that the most striking increase in nitrogen balance, with an increase in caloric intake, occurred when the intake increased from 0 to 40 kcal/kg ideal body weight, and that increasing calories above 40 kcal/kg ideal body weight did not appreciably increase nitrogen balance (Figure 5). This figure corresponds to the theoretical maximum calculated on the basis of a 60% increase in the metabolic rate referred to earlier in this chapter and amounts to an intake of 2800 kcal/day in the 70 kg man. This caloric intake will also meet the requirements for energy consumption noted in the injured-septic patient by both Askanazi et al. and Roulet et al.

IV. ELECTROLYTES AND TRACE ELEMENTS

The importance of fluid and electrolyte replacement for promoting tissue perfusion and ionic equilibrium is self-evident. In addition, the processes of malnutrition and refeeding are both associated with major changes in electrolyte balance. With protein-calorie malnutrition there is loss of the intracellular ions potassium, magnesium, and phosphorus, together with a gain in sodium and water. On refeeding it is necessary to give potassium, magnesium and phosphorus, and zinc to ensure optimum nitrogen retention. The sodium balance may initially become markedly positive and cause edema during refeeding, particularly with carbohydrate (see reference by Veverbrants and Arky).

A. Sodium

Sodium is the principle extracellular ion and normally the major part of body sodium is in the extracellular fluid where it is present in a concentration of about 140 mmol/ℓ. Since extracellular fluid amounts to about 20% of body weight, the total amount of sodium in the extracellular fluid (ECF) is about 1960 mmol. In contrast, the sodium concentration in the intracellular fluid is only about 5 mmol/ℓ. Since intracellular fluid amounts to 50% of body weight therefore the total intracellular sodium is only 175 mmol. Therefore total body sodium in the adult will average 30 mmol/kg body weight. In malnourished children it has been shown that the total body sodium is increased. In surgical patients, the majority of whom were below 85% of ideal body weight, Hill et al. found that the total body sodium was 2789 mmol in patients who on the average weighed 51.7 kg, which on recalculation amounts to 53.9 mmol/kg — almost double the expected 30 mmol/kg. On giving TPN this rose to 3038 mmol, and since they also gained 2.6 kg body weight on the average, the total body sodium increased to 56 mmol/kg although the authors did not find this rise statistically significant. In another study with oral diets Veverbrants et al. indicated that carbohydrate feeding induces sodium retention not seen when isocaloric amounts of fat or protein are fed. Hence it is clear that in malnourished patients great care should be taken to prevent salt and water overload, especially in elderly subjects and those with cardiopulmonary disease. The more malnourished the patient the greater this danger, especially when large amounts of carbohydrate are being infused. The obvious clinical manifestation of fluid overload is the appearance of "refeeding edema".

Recommendations: In the average patient about 100 to 120 mmol/day of sodium can be given. This value is supplemented to cover abnormal losses via the GI tract. In severely malnourished patients, and those with cardiopulmonary disease, the sodium intake should be restricted to 50 to 60 mmol/day and the amount infused gradually increased as the patient tolerates the fluid load. In this context it has been shown that potassium and magnesium deficiency may prevent the excretion of free water resulting in hyponatremia even though total body sodium is increased — because total body water is increased even more. Under these circumstances appropriate ion replacement and the use of diuretics and water restriction may have to be practiced. In any case, severe edema must be avoided in these very sick and malnourished patients.

B. Potassium

Potassium is the main intracellular cation and its concentration in cells amounts to about 140 mmol/ℓ. On the basis of body water distribution indicated above (under Sodium) it can be calculated that the 70-kg man has 4900 mmol of intracellular potassium. Based on an average extracellular potassium of 4 mmol/ℓ, the extracellular distribution of potassium amounts to 56 mmol giving a total body potassium of 71 mmol/kg. In children with protein-calorie malnutrition the total body potassium is markedly reduced and this deficiency is not reflected in plasma levels. In malnourished surgical patients recalculation of the data by Hill et al. for the total body showed that potassium was only 39 mmol/kg and TPN increased it to 43 mmol/kg without increasing total body nitrogen. Mernagh et al. also noted that body potassium was disproportionately reduced in malnourished patients compared with nitrogen and was repleted by short-term (< 30 days) TPN without altering total body nitrogen. In contrast, in this study it was noted that long-term TPN increased both nitrogen and potassium. Hence, it should be recognized that potassium deficiency is an integral part of malnutrition and needs repletion apart from nitrogen. Rudman et al. showed that a positive nitrogen balance during glucose-based TPN did not occur unless potassium was also given.

Recommendations: The next question is the potassium requirement during parenteral nutrition. To answer this question three facts need to be considered. First, that glucose infusions will increase the need for potassium, second, that about 3 meq of K^+ are retained with each gram of nitrogen and third, that based on the data of both Hill and Mernagh, the total deficit of potassium may amount to between 800 and 900 mmol in the 50 to 70 kg adult. The infusion of about 80 to 120 meq/day will aid in replenishing stores and meeting daily needs.

C. Magnesium

Magnesium is the next most abundant intracellular ion, after potassium. Intracellularly it is bound to protein and is necessary for a number of vital cellular functions which include membrane and mitochondrial integrity, enzyme activation (including that of ATPase and adenyl cyclase) and also for the synthesis and stability of nuclear DNA. It is also necessary for control of neuromuscular excitability. The average concentration in plasma is 1.0 mmol/ℓ or 2.0 meq/ℓ and based on a normal extracellular volume a 70-kg man has about 14 mmol extracellularly. This constitutes only about 3% of the total, the intracellular magnesium amounting to about 500 mmol. In protein-calorie malnutrition magnesium deficiency has been documented and in surgical patients Freeman found that increasing magnesium intake up to about 15 mmol/day improved nitrogen balance. In malnourished patients on home TPN Jeejeebhoy and colleagues have documented marked positive magnesium balance and noted that 15 to 17 mmol/day was required to maintain balance and normal serum levels over the long term, figures that agree with the finding of Freeman. Furthermore it should be noted that in patients with a short bowel the needs were especially high due to losses from diarrhea and stomal drainage.

Recommendations: Magnesium should always be added to parenteral nutrition regimens. At least 12 to 15 mmol/day should be given and, in addition, extra should be added to cover losses from GI secretions.

D. Phosphorus

Phosphorus is the major intracellular anion. In the cell, among other functions, it is a part of a major buffer system, of energy storing nucleotides (ATP), of membranes as phospholipids, and of oxygen transfer systems in the form of erythrocyte 2,3 diphosphoglycerate (2,3 DPG). The total body phosphorus outside the skeleton amounts to 6000 mmol in the 70-kg man.

Since the serum levels are only about 1.2 mmol/ℓ, extracellular phosphorus amounts to only 17 mmol in the 70-kg man, showing that the major part of this element is intracellular. Data recalculated from Hill et al. showed that the total body phosphorus in malnourished surgical patients was only 281 mmol/kg compared with the normal of 374 mmol/kg. If it is assumed that skeletal phosphorus is unchanged in acute malnutrition, then such a difference would suggest that intracellular phosphorus was proportionately even more depleted.

The serum phosphorus falls rapidly during TPN and the literature is full of reports about the serious effects of hypophosphatemia, which in some cases proved to be lethal. Silvas showed that during parenteral nutrition where all nonprotein calories were given as glucose, serum phosphorus may drop precipitously followed by the development of tremors, impaired mentation, paresthesias, muscular weakness, convulsions, and coma. These symptoms usually occur when glucose-based parenteral nutrition provides all caloric needs to malnourished patients. However, in our experience this syndrome has not been seen in any patient infused with a glucose-lipid-based nutrient regimen, and hence such a dual substrate system is safer in severely malnourished patients. In addition, Rudman et al. showed that a positive nitrogen balance was associated with phosphorus retention and hence this element is of importance in promoting anabolism.

Recommendations: The total needs for phosphorus amount to about 14 to 16 mmol/day when a glucose-lipid source of nonprotein energy is being given. These requirements are increased when glucose alone is given as a source of energy. This is partly due to the fact that lipid emulsions have phospholipids which act as an additional source of phosphorus and also because the high insulin levels associated with the glucose system increase cellular uptake of phosphorus.

E. Zinc

Zinc is the most abundant of all the trace elements, amounting to between 22 and 35 mmol of which 99% is intracellular and only 1% extracellular. It is therefore not surprising that plasma zinc is a poor indicator of zinc stores or of zinc balance. Zinc is an integral part of about 60 different enzymes and is necessary for RNA, DNA, and protein syntheses.

Clinical zinc deficiency during parenteral nutrition was noted by both Kay et al. and Arakawa et al., presenting with a scaly and pustular rash starting around the mouth and over the articular prominences and spreading over the whole body. In addition there is erythema of the palms and soles and alopecia. Superficial skin infections are commonly noted and resistant to treatment. In a controlled study Wolman et al. showed that zinc requirements in patients receiving TPN were dependent on the source and volume of GI tract losses. Patients without diarrhea required about 2.5 to 3.0 mg of zinc per day. In contrast, the presence of diarrhea, ostomy losses, and intestinal drainage increased the requirements. It was estimated that the losses of zinc in upper intestinal fluid were about 10 to 12 mg/ℓ, but rose to between 17 and 20 mg/ℓ in diarrheal stool. In addition about 2 to 5 mg of zinc per day was lost in the urine. Urinary zinc losses were increased in the injured-septic patient. In view of the relationship of the volume of intestinal contents lost and zinc losses, zinc requirements can be simply estimated by measuring the volume of losses from the GI tract and calculating zinc replacement to meet these losses. In addition, it is also necessary to give extra zinc to make up for any deficiencies. In the context of renal failure one may note that since urinary losses were found to be relatively modest and independent of the infusion rate, zinc infusions may be given even in the presence of renal failure.

The above study also showed that once TPN was started, the level of plasma zinc did not reflect zinc balance and was a poor indicator of deficiency. Furthermore, in this study a positive zinc balance was associated with an increase in nitrogen retention and enhanced insulin secretion; in the absence of a positive zinc balance, glucose metabolism was found to be abnormal.

Recommendations: In patients without diarrhea or fistula losses, about 2.5 mg of zinc should be added to the parenteral infusates. With diarrhea, the amount added would depend on the volume of diarrheal stool passed by the patient. On the average 0.26 mmol should be added per liter of diarrheal fluid lost in the NPO state. With small bowel fistula losses this value should be reduced to 0.15 mmol/ℓ of fistula fluid in the NPO state.

F. Copper

Copper is an integral part of many oxidative enzymes such as ceruloplasmin, cytochrome *c* oxidase, and lysyl oxidase. Ceruloplasmin aids the oxidation of ferrous iron in tissue stores to the ferric form, to enable it to be transported by transferrin. Lysyl oxidase aids the oxidation of the C-terminal carbon of lysine in collagen, which is a prerequisite for cross-linkage. Therefore copper deficiency results in anemia with an iron deficiency picture, leukopenia, and, in children, bone disease. Altogether there are 1.3 to 2.0 mmol of copper in the human body, mainly intracellular, residing in muscle, bone, liver, and brain. In the circulation copper is mainly bound to ceruloplasmin and, in general, plasma copper levels reflect those of this protein. Copper is excreted almost entirely in the bile, associated closely with taurodeoxycholate.

In order to determine the requirements of copper during TPN, Shike et al. performed a controlled study during which patients received 0, 0.8, and 1.6 mg of added copper to their TPN. Each level of copper was infused for 1 week at a time, and the order in which the different doses of copper were infused was randomized to ensure that changes in the patient's clinical condition during the hospital stay did not bias the observed results. The data showed that as with zinc, the plasma copper levels did not reflect the balance. The copper losses from the GI tract were independent of the volume of GI contents excreted, a finding which contrasts with that observed with zinc. However, the losses were significantly lower in patients with normal bowel habits compared to those with diarrhea. Furthermore, GI losses were found to decrease with the onset of jaundice and therefore jaundiced patients should not be given copper.

Recommendations: In the absence of diarrhea about 0.2 to 0.3 mg of copper should be added to TPN solutions per day. With diarrhea this should be increased to 0.5 mg/day. In jaundiced patients, or those with liver dysfunction, copper should be withheld.

G. Chromium

This element has been found to be necessary for glucose homeostasis. In its absence insulin-resistant diabetes and neurological changes have been noted during TPN. About 10 to 20 μg/day is required.

H. Selenium

Deficiency of selenium was demonstrated during TPN by Van Rij et al. The syndrome of deficiency was characterized by low levels of selenium in the erythrocytes together with muscle dysfunction responding to selenium administration.

I. Vitamins

Vitamins are essential nutrients which are active in minute quantities. While it is obvious that these substances have to be given in any regimen of TPN to avoid deficiency, with the exception of vitamin D, the optimum dose and frequency of administration has not been studied in detail in patients receiving TPN. The currently available studies have been simple observations of plasma or blood levels during a given regimen.

1. Vitamin A

This vitamin is essential for the integrity of epithelial surfaces, synthesis of retinal pigments, and is also of importance in protecting against infection. It is fat soluble and is stored in the liver. In patients on long-term TPN, 2500 IU maintained normal circulating levels for periods of several years. When all vitamin A was withdrawn for 6 months these patients did not show a fall in plasma levels, indicating that 2500 IU/day had maintained adequate stores. Furthermore, despite serious episodes of sepsis in three of these patients, there was no fall in plasma vitamin A levels and no difficulty in recovering from infection. Therefore, 2500 IU is recommended as the daily dose of this vitamin by the parenteral route.

2. Vitamin D

In long-term TPN, 250 IU was found to maintain normal plasma levels of 1,25-hydroxycholecalciferol, a metabolite of this vitamin. However observations by Shike et al. indicated that patients on long-term TPN developed a syndrome of hypercalciuria, intermittent hypercalcemia, osteomalacia with bone pains, and fracture of the axial skeleton. In the short-term TPN patient the hypercalcemia was associated with pancreatitis. All these changes were reversed and clinical healing of fractures occurred by simply withdrawing the vitamin D. This improvement occurred despite the fact that an active metabolite of this vitamin, 1,25-dihydroxycholecalciferol, was found to be low in these patients. Hence it is clear that during TPN intravenous vitamin D is undesirable for reasons not entirely clear at this time. Furthermore, when observed over one year, the withdrawal of vitamin D did not result in subnormal plasma levels of the 25-monohydroxy derivative. On the basis of these findings no intravenous vitamin D is recommended and instead long-term TPN patients are encouraged to expose themselves to sunlight to maintain normal D levels from natural sources.

3. Vitamin E

This vitamin is an antioxidant and its necessity in human nutrition is only clearly documented in infants, in whom deficiency causes hemolytic anemia. In adults the need for this vitamin is controversial, but in a patient on long-term TPN Howard and colleagues noted neurological changes which improved specifically after the administration of vitamin E. Furthermore, the requirement for vitamin E is believed to be enhanced when polyunsaturated fatty acids are added to the diet. Since lipid emulsions contain polyunsaturated fatty acids the need for this vitamin may be increased during the administration of lipid emulsions. Fortunately these emulsions contain significant amounts of this vitamin in the recommended proportions of 0.4 IU/g fatty acid, and indeed, in patients on long-term TPN receiving 50 g of lipid a day the level of vitamin E was almost within the normal range. It is recommended that a total of 30 to 50 IU be infused in patients on TPN, of which a part may come from lipid emulsions and part from added vitamin.

4. Vitamin K

This vitamin is required for the synthesis of four factors necessary for coagulation. While there are no data about the precise needs for this vitamin, nevertheless administering 10 mg of Synkavite® (a water-soluble analogue of this vitamin) per week was shown to maintain normal coagulation parameters in patients on long-term TPN observed over several months.

5. Thiamin

Thiamin is an integral part of the cocarboxylase enzyme complex which is necessary for the metabolism of alpha-keto acids such as pyruvate. Cells such as neurones, which depend exclusively on carbohydrates as an energy substrate, need this vitamin and are especially vulnerable to thiamin deficiency. The requirements amount to about 1 mg/1000 kcal of energy intake. Based on observations of the thiamin dependent enzyme activity in such patients, Kishi et al. found that 5 mg/day meets the requirement in patients receiving short-term TPN. In patients receiving long-term TPN, 16 mg of thiamin per day resulted in blood levels which were three times higher than the mean normal value for this vitamin. Hence it is considered that 5 mg/day should be sufficient for patients receiving TPN.

6. Riboflavin

This vitamin is a component of the coenzymes flavin mononucleotide and flavin adenine dinucleotide, which are necessary for proton transfer in redox systems. Riboflavin deficiency causes photophobia, glossitis, chelosis, and pruritus of the skin, especially with inflammation around the anogenital area. In patients receiving TPN, 5 mg/day is recommended as being a safe amount in order to avoid deficiency.

7. Niacin

Niacin is a component of nicotine adenine dinucleotide (NAD) and its phosphate (NADP) which are necessary for several dehydrogenation reactions required for carbohydrate and protein metabolism and for cell respiration. In deficiency states the syndrome of pellagra develops, characterized by dark red erythema on the exposed areas of the body with pigmentation and cracking of the skin, glossitis, stomatitis, and diarrhea. In addition, neurological symptoms occur with delirium and confusion. On the basis of observed circulating levels of this vitamin in long-term TPN patients it is recommended that about 50 mg/day be considered a safe intake to avoid deficiency.

8. Pantothenate

Pantothenic acid is a component of coenzyme A. The deficiency of this vitamin is poorly characterized in man but appears to result in fatigue, mental disturbances, paresthesia, and epigastric discomfort. Studies of blood levels of this vitamin in patients on long-term TPN suggests that 15 mg/day is a suitable intake.

9. Pyridoxine

This vitamin in a coenzyme in many reactions concerned with amino acid metabolism including decarboxylation, transamination, and dehydroxylation. In the adult human the symptoms of deficiency are not very specific, consisting of dermatitis, intertrigo, seborrhea, irritability, somnolence, and neuropathy. An intake of 5 mg/day is sufficient to maintain normal blood levels of this vitamin in patients on long-term TPN.

10. Folic Acid and Vitamin B_{12}

These vitamins are important for the synthesis of nucleic acids. Deficiency results in megaloblastic anemia and glossitis. Severe vitamin B_{12} deficiency will result in neuromyelopathy. The recommended intake parenterally is 600 μg of folic acid and 12 μg of vitamin B_{12} per day.

11. Vitamin C

Vitamin C is a strong reducing agent and is required for redox reactions, collagen synthesis, and normal immune functions. In deficiency states the syndrome of scurvy occurs with perifollicular hemorrhages in the skin, gingivitis, and infections. In children, subperiosteal hemorrhages may occur. About 300 to 500 mg/day will maintain normal-to-high blood levels of this vitamin in patients receiving TPN.

12. Biotin

The need for this vitamin in patients receiving TPN is not obvious in adults. In long-term TPN, despite very low biotin levels, we found no evidence of clinical deficiency. Recently infants on TPN have shown no signs of deficiency. 300 mg/day can be given.

V. ESSENTIAL FATTY ACID DEFICIENCY

In the human, linoleic acid — a fatty acid with 18 carbon atoms and 2 unsaturated (double) bonds — is not synthesized *de novo* from other substrates and has to be provided in the diet. Hence it is called an essential fatty acid (EFA). This fatty acid has another distinguishing characteristic, namely the omega number. This number is the number of carbon atoms between the methyl end and the first double bond; in the case of linoleic acid it happens to be 6. Another fatty acid, linolenic acid, has been suggested as being essential in the human infant.Linoleic acid is elongated into a 20-carbon fatty acid with 4 double bonds called arachidonic acid which is, in turn, a precursor of prostaglandins. In addition, linoleic acid is an essential constituent of membranes where it contributes to membrane pliability and fluidity. When there is a deficiency of linoleic acid then the chain elongating enzyme system will utilize oleic acid as a substrate and elongate it to another 20-carbon fatty acid with 3 double bonds called eicosatrienoic acid. This fatty acid is not present in normals and its appearance suggests the presence of EFA deficiency. Since eicosatrienoic acid has three double bonds it is a triene. In contrast arachidonic acid has four double bonds and is a tetraene. From the above considerations a rise in the triene/tetraene ratio suggests EFA deficiency, and this ratio is so used.

Normally the adult has plentiful fat stores available to supply EFA, but during glucose-based parenteral nutrition the high insulin level generated by the infusion of hypertonic glucose suppresses the mobilization of essential fatty acids from fat stores, causing the early development of EFA deficiency. Hence in this situation lipid infusions have to be given to meet these requirements. Dose-response studies performed in patients receiving long-term TPN showed that the least amount of EFA which maintained an optimum ratio of linoleate/oleate in plasma phospholipids was about 7% of calories. This proportion of EFA to total calories is provided by infusing 350 mℓ of a 10% fat emulsion per day.

On the other hand, as indicated earlier, with the increasing use of fat as a source of calories the whole question of EFA deficiency is resolved.

Chapter 4

EVALUATION OF THERAPY

Khursheed N. Jeejeebhoy

TABLE OF CONTENTS

I. INTRODUCTION

The recognition that malnutrition is widespread in hospitals, together with the development of techniques to treat such malnutrition, led to the appreciation that a precise definition of the features of malnutrition in its early stages and of methods to quantitate the nutritional status of the patient were lacking. In an effort to assess a patient's nutritional status using relatively simple techniques, Blackburn et al. advocated a series of observations designed to define the anthropomorphic, biochemical, and immunological status of the patient and called it the nutritional assessment profile. However, critical evaluation of the ability of this profile to discriminate early malnutrition has brought out several nuances which suggest that such a profile may not be specific for malnutrition per se, and that the total profile (in contrast to specific measurements) may not be necessary to predict morbidity and mortality due to malnutrition. Furthermore, malnutrition and refeeding have complex effects on body composition which cannot be defined by simple means. These aspects will now be considered in turn.

II. ROLE OF HISTORY AND PHYSICAL EXAMINATION

In the assessment of an individual patient there are two considerations — first, whether the patient has become malnourished and second, whether the conditions causing malnutrition are likely to continue in the future. For both these purposes the history and physical examination are likely to provide more information than static anthropometric measurements. The concept of malnutrition is a difficult one to define in all its grades because we do not know as yet the changes in body composition which result in a dangerous degree of functional impairment. For example, long-distance runners are often thinner and have significantly lower weights that the so-called normal mean, but they do not have any functional disability. Hence each person has to be compared to both his or her own norms and to absolute values. Here a dietary history together with comparative data of the pre- and postmorbid states will reveal whether there has been significant development of malnutrition and also whether the intake of food has been altered by disease in such a way as to be likely to cause malnutrition. In addition, the physician should not wait for malnutrition to occur but should take prophylactic action by predicting the course and nature of the illness and its impact on intake, need, and utilization of nutrients. These objectives can be met by careful attention to the clinical status of the patient rather than the static anthropometric data. In a controlled study Baker et al. compared the ability of the clinician to discern malnutrition by history and physical examination in comparison to anthropometry, biochemistry, and whole-body composition studies. The clinical examination was performed blind by two clinicians who categorized the patients as being within normal limits, moderately malnourished, or severely malnourished. The outcome was that the two observers agreed in 81% of instances, showing the reproducibility of the clinical examination. Furthermore their judgment coincided with inferences drawn from objective parameters derived from anthropometry and body composition studies. This study indicated clearly that clinical assessment could be an effective means of judging the nutritional status of the patient.

III. ANTHROPOMETRIC MEASUREMENTS

A. Weight/Height

The weight of an individual is approximately related to height, and studies by life insurance companies have revealed that for each height there is a range of body weight which is associated with the lowest mortality. This range is the so-called ideal body weight. However inspection of a table of so-called ideal body weights shows a range of about 20% around the mean, depending on a judgment of "frame size". This makes it very difficult to decide whether individual patients have lost >15% of body weight, which is a commonly used indicator of impaired nutrition. A more reliable index is a fall in body weight of the same magnitude from the usual body weight. A further problem arises in the case of chronically malnourished patients who may then have a superadded acute insult with a further fall in body weight. In such patients, reference to the usual body weight may underestimate the significance of weight loss. Edema also may result in underestimating weight loss. Clearly, a careful history together with current weight measurements is needed in order to assess weight loss.

B. Skinfold Thickness

Skinfold thickness is a means of estimating body fat. The techniques range from a simple measurement of skinfold thickness over the triceps midway between the shoulder and elbow, to a formal evaluation of body density by multiple skinfold measurements together with an estimate of body frame by the measurement of wrist size and waist circumference. These measurements allow an estimate of body fat as a percentage of body weight. Then by using this value for body fat, some measure of lean weight can be obtained from the relationship:

Lean body weight = (100 − % body fat) × total body weight/100

We have used the following formulas to estimate body fat (after Sinning et al.):

Males:

1. % body fat = 8.71 + [0.49 × waist circumference (cm)] + [0.49 × pectoral skinfold (mm)] − [6.36 × wrist diameter (cm)].
2. Anthropometrically measured lean body mass (AMLBM) = wt × (1 − % body fat /100).

Females:

1. AMLBM = 8.99 + [0.73 × weight (kg)] + [3.79 × wrist diameter (cm)] − [0.16 × waist circumference (cm)] − [0.25 − hip circumference (cm)] + [0.43 × forearm diameter (cm)]
2. % body fat = (1 − AMLBM/wt) × 100.

Unfortunately these methods tend to be accurate only in normals and not so accurate in wasted or obese subjects. Nevertheless, using the Sinning formulas the author has noted that the anthropometrically measured lean body mass (AMLBM) correlated well with total body nitrogen in the normal subject. The regression of the relationship suggested that 3.7% of lean body weight was nitrogen, which is close to the figure of 3% estimated by carcass analysis. However after intravenous feeding in patients, we found that AMLBM was a better measure of body potassium than of body nitrogen. Since total body potassium reflects intracellular water content, AMLBM seems to be a better measure of body cell volume than of body protein content.

C. Mid-Arm Circumference

This is estimated by measuring the circumference of the arm midway between the shoulder and elbow, and correcting the measurement for subcutaneous fat. Mid-arm circumference (cm) = measured circumference − (0.314 × triceps skinfold thickness).

This measurement is compared with normal values and is used to estimate body protein stores. However the assumptions upon which this estimate is based are only approximately true, and we have found that this measurement correlated very poorly with total body nitrogen.

IV. CREATININE-HEIGHT INDEX

The 24-hr creatinine excretion in the urine is dependent on the muscle mass, provided that renal function is normal. The creatinine excretion in an individual can be defined in terms of an expected value for a given height by reference to tables (Table 1):

Creatinine-height index = actual urinary creatinine/ideal urinary creatinine

The result is an index of body muscle mass as a percentage of normal. In our studies this index correlated well with total body nitrogen but not with total body potassium. Hence this index does seem to reflect the mass of body protein.

V. SIGNIFICANCE OF ANTHROPOMETRIC MEASUREMENTS AND THE CREATININE-HEIGHT INDEX

The above measurements are designed to assess whether there has been significant wasting of total body mass, fat stores, and muscle bulk, respectively. There are two main difficulties in using these anthropometric parameters. First, there is the question of what is normal for an individual because of wide variation in so-called "normal" values depending upon habitus, exercise, and diet. Second, the question is not only one of absolute numbers alone but what these figures mean in terms of risk to vital body function. Clearly, any objective parameter of malnutrition can only be interpreted in regard to its clinical significance if it can be shown to be associated with functional changes likely to endanger the patient or to hinder recovery. This correlation has not been defined or studied and hence the ultimate significance of moderate changes in these parameters needs further studies.

VI. PLASMA PROTEIN MEASUREMENTS

1. Serum albumin
2. Serum transferrin (TIBC $\times$ 0.8 $-$ 43)

These proteins are synthesized by the liver and the rates of synthesis of both these proteins are controlled by nutritional factors. Hence the levels of both these proteins in the plasma have been used as an index of the effect of the nutritional status of the patient on visceral protein synthesis. Unfortunately the plasma level of any protein is a result of a balance between synthesis and catabolism, and therefore a reduced level may equally be due to reduced synthesis or increased catabolism. In GI diseases it was shown years ago that the reduced plasma albumin levels observed were mainly due to increased catabolism resulting from GI protein loss rather than reduced synthesis. Under these circumstances the low circulating levels may not respond to nutrient supplementation alone, but improve when the basic disease undergoes remission. For example, the hypoalbuminemia of Crohn's disease has clearly been shown to be due to protein loss and it is not surprising if, in patients with this disease, simply giving nutrient may only increase total body protein without raising plasma protein levels. As an extension of the above concept it should be recognized that plasma proteins are not a good index of nutrition in patients with GI disease without excluding GI protein loss.

In addition to losses of plasma proteins via the bowel, such loss may occur from burn wounds, extensive ulcers, and in renal disease. In the presence of such losses it is not possible to use the level of these proteins as an index of malnutrition.

Table 1
REFERENCE TABLE FOR URINARY CREATININE/
BODY HEIGHT FOR PERSONS OF IDEAL WEIGHT
FOR THEIR HEIGHT

FOR MEN
(Creatinine Coefficient: 23 mg/kg/body weight)

Height			Ideal weight (medium frame)		Total creatinine (mg/24 hr)	Total creatinine (mg/cm body height/24 hr)
ft	in.	cm	lb	kg		
5	2	157.5	124	56	1288	8.17
5	3	160	127	56.6	1325	8.28
5	4	162.6	130	59.1	1359	8.36
5	5	165.1	133	60.3	1386	8.40
5	6	167.6	137	62	1426	8.51
5	7	170.2	141	63.8	1467	8.62
5	8	172.7	145	65.8	1513	8.76
5	9	175.3	149	67.6	1555	8.86
5	10	177.8	153	69.4	1596	8.98
5	11	180.3	158	71.4	1642	8.11
6	0	182.9	162	73.5	1691	9.24
6	1	185.4	167	75.6	1739	9.38
6	2	188	171	77.6	1785	9.49
6	3	190.5	176	79.6	1831	9.61
6	4	193	181	82.2	1891	9.80

FOR WOMEN
(Creatinine coefficient: 18 mg/kg/ body weight)

4	10	147.3	101.5	46.1	830	5.63
4	11	149.9	104	47.3	851	5.68
5	0	152.4	107	48.6	875	5.74
5	1	154.9	110	50	900	5.81
5	2	157.5	113	51.4	925	5.87
5	3	160	116	52.7	949	5.93
5	4	162.6	119.5	54.3	977	6.01
5	5	165.1	123	55.9	1006	6.09
5	6	167.6	127.5	58	1044	6.23
5	7	170.2	131.5	59.8	1076	6.32
5	8	172.7	135.5	61.6	1109	6.42
5	9	175.3	139.5	63.4	1141	6.51
5	10	177.8	143.5	65.2	1174	6.60
5	11	180.3	147.5	67	1206	6.69
6	0	182.9	151.5	68.9	1240	6.78

From Blackburn, G. L. et al., Manual for nutritional metabolic assessment of the hospitalized patient, presented in a poster session at the 62nd Ann. Clin. Cong. Am. Coll. Surg., Chicago, 1976. With permission.

VII. PARAMETERS OF IMMUNITY

1. Delayed cutaneous hypersensitivity
2. Lymphocyte count

There has been considerable evidence in kwashiorkor patients that protein-calorie malnutrition may cause anergy. These observations have led to similar studies in seriously ill patients and resulted in the observation by Meakins et al. that anergy to cutaneously injected recall antigens was associated with a very high mortality. Furthermore it was claimed that TPN with reversal of anergy improved survival — the belief being that reversal of protein-calorie malnutrition was the main therapeutic factor. In contrast Golden et al. showed that even in the presence of protein-calorie malnutrition, the provision of zinc reversed anergy, casting doubt on the role of protein and calories per se being the cause of the anergy. This study also highlighted a hitherto unrecognized role for micronutrients in aiding body immunity. In studies of 52 patients by Baker et al., 55% of the severely malnourished patients were anergic, however none died in the hospital and all were followed to the point of discharge from the hospital. In addition, 17% of the patients without any evidence of malnutrition were anergic. Hence anergy cannot predict the outcome in any single patient and is not an index of protein-calorie malnutrition, per se. This concept is supported by several recent publications including those of Meakins et al., McLaughlin et al., Wang et al., and Miller, suggesting that in sick patients many factors other than malnutrition may affect immunity and skin reactivity to antigens. These factors include fever, sepsis, drugs, tumors, and shock together with circulating inhibitors of lymphocyte function — altogether making delayed cutaneous hypersensitivity a poor index of malnutrition or of the risk of morbidity and mortality resulting from malnutrition per se, rather than from the disease.

VIII. MEASUREMENT OF BODY COMPOSITION BY NONINVASIVE METHODS

A. Total Body Potassium

Potassium is the principal intracellular cation and as such the total body potassium (TBK) reflects the volume of intracellular water because the osmolarity of living cells is controlled within narrow limits. TBK may be measured by counting naturally occurring ^{40}K in a whole-body counter or by isotope dilution using ^{42}K. Among tissues in general, muscle is especially rich in potassium by comparison with nitrogen and therefore it is not surprising that males have a higher ratio of TBK to total body nitrogen (TBN). In the normal person there is good correlation between TBK and TBN, but this relationship becomes disturbed by malnutrition. During acute malnutrition there is a marked fall in TBK, which in relation to the loss of TBN is greater in males than in females. After short-term (< 30 days) TPN, all patients Jeejeebhoy and associates studied who gained weight had a rise of TBK but not necessarily of TBN. In contrast, after long-term TPN, TBK and TBN rose proportionately, showing that the initial effect of TPN is to restore potassium and intracellular water and not nitrogen as has been believed.

B. Total Body Nitrogen

This is measured by two different methods. The first is prompt gamma emission (PGE). This reaction is the prompt emission of gamma radiation when a neutron is captured by a nitrogen molecule. Simultaneously, a gamma ray of a different energy is released by hydrogen when bombarded by neutrons. The latter event can be used as an internal standard for the absorption of the neutron flux by body tissues. Hydrogen can be used as an internal standard because it is present in almost the same concentration in all soft tissues and it is therefore possible to calculate the nitrogen-to-hydrogen count ratio and to derive the mass of nitrogen from the known proportion of hydrogen to body weight. Neutron bombardment of the body and the gamma emissions from it are measured concurrently. This method has been validated by comparing the results of PGE with simultaneously performed chemical analysis of porcine carcass.

The other method is neutron activation — where neutron bombardment of the patient creates a short-lived isotope of nitrogen which can be subsequently counted in a whole-body counter.

The PGE method has the advantage that it gives much less radiation, equivalent to only that of a chest X-ray. On the other hand, neutron activation allows multielement analysis which is not possible with the PGE method.

C. Total Body Water

Lean body tissue is composed largely of water; in contrast there is little water in adipose tissue. Hence measurements of total body water reflect the volume of lean tissues if the extracellular fluid compartment does not expand.

IX. RESULTS OF NUTRIENT SUPPORT ON BODY COMPOSITION

A. Central Venous Hyperalimentation with Glucose

Yeung and colleagues have shown that with central venous hyperalimentation the massive weight gain was mainly due to an increase in body fat and water, and not nitrogen. Despite the lack of increase in body nitrogen there was a substantial rise in body potassium, showing that the increase in body potassium under the influence of glucose and insulin did not reflect nitrogen anabolism. Shizgal and co-workers claimed, on the basis of body potassium studies, that only glucose infusions increased "body cell mass", the assumption being that an increase in body potassium implies a specific increase in the protein of the cell mass. Inasmuch as potassium reflects intracellular volume, their observation suggests nothing more than the fact that glucose with insulin drives potassium into cells. It does not imply an increase in nitrogen retention for reasons given earlier.

B. Dual Substrate (Lipid and Glucose) Infusions

When both glucose and lipid are used there is a disproportionate increase in body nitrogen compared with body potassium or body weight. These findings, together with those given above, suggest that a dual substrate input is the best way of increasing body nitrogen.

C. Hypocaloric Protein Infusions and Defined Formula Diets

Both hypocaloric infusions of amino acids alone and the feeding of defined formula diets, failed to achieve an increase of body nitrogen during a three-week study. Thus these modalities only maintained total body nitrogen and did not bring about an increase in this parameter. Of interest is that adding glucose calories did not add to the nitrogen retention (see reference by Yeung et al.).

READING LIST

(Chapters 1 to 4)

Anderson, G. H., Bryan, H., Jeejeebhoy, K. N., and Corey, P., Dose-response relationships between amino acid intake and blood levels in newborn infants, *Am. J. Clin. Nutr.*, 30, 1110, 1977.

Anderson, G. H., Patel, D. G., and Jeejeebhoy, K. N., Design and evaluation by nitrogen balance and blood aminograms of an amino acid mixture for total parenteral nutrition of adults with gastrointestinal disease, *J. Clin. Invest.*, 53, 904, 1974.

Arakawa, T., Tamura, T., Igarashi, Y., Suzuki, H., and Sandstead, H. H., Zinc deficiency in two infants during total parenteral alimentation for diarrhea, *Am. J. Clin. Nutr.*, 29, 197, 1976.

Askanazi, J., Carpentier, Y. A., Elwyn, D. H., Nordenstrom, J., Jeevanandam, M., Rosenbaum, S. H., Gump, F. E., and Kinney, J. M., Influence of total parenteral nutrition on fuel utilization in injury and sepsis, *Ann. Surg.*, 191, 40, 1980.

Askanazi, J., Rosenbaum, S. H., Hyman, A. T., Silverberg, P. A., Milic-Emili, J., and Kinney, J. M., Respiratory changes induced by the large glucose loads of total parenteral nutrition, *JAMA*, 243, 1444, 1980.

Baker, A., Wagonfeld, J. B., and Rosenberg, I. H., Intractable sprue successfully treated by total parenteral alimentation (TPA) followed by selective protein-restricted diet, *Gastroenterology*, 68 (Abstr.), 1054, 1975.

Bark, S., Holm, I., Hakansson, I., and Wretlind, A., Nitrogen-sparing effect of fat emulsion compared with glucose in the post-operative period, *Acta Chir. Scand.*, 142, 423, 1976.

Blackburn, G. L., Bistrian, B. R., Maini, B. S., Benotti, P., Bothe, A., Gibbons, G., and Smith, N. F., Manual for Nutritional Metabolic Assessment of the Hospitalized Patient, paper presented at the 62nd Ann. Clin. Cong. Am. Coll. Surg., Chicago, 1976.

Blackett, R. L. and Hill, G. L., Post-operative external small bowel fistulas: a study of a consecutive series of patients treated with intravenous hyperalimentation, *Br. J. Surg.*, 65, 775, 1978.

Blackburn, G. L., Flatt, J. P., Clowes, G. H. A., and O'Donnell, T. E., Peripheral intravenous feeding with isotonic amino acid solutions, *Am. J. Surg.*, 125, 447, 1973.

Bozzetti, F., Parenteral nutrition in surgical patients, *Surg. Gynecol. Obstet.*, 142, 16, 1976.

Burke, J. F., Wolfe, R. R., Mullany, C. J., Mathews, D. W., and Bier, D. M., Glucose requirements following burn injury. Parameters of optimal glucose infusion and possible hepatic and respiratory abnormalities following excessive glucose intake, *Ann. Surg.*, 190, 274, 1979.

Buzby, G. P., Mullen, J. L., Matthews, D. C., Hobbs, C. L., and Rosato, E. F., Prognostic nutritional index in gastrointestinal surgery, *Am. J. Surg.*, 139, 160, 1980.

Calloway, D.H. and Spector, H., Nitrogen balance as related to calorie and protein intake in active young men, *Am. J. Clin. Nutr.*, 2, 405, 1954.

Collins, J. P., Oxby, C. B., and Hill, G. L., Intravenous amino acids and intravenous hyperalimentation as protein-sparing therapy after major surgery. A controlled clinical trial, *Lancet*, 1, 788, 1978.

Cummins, G. E., Grace, A. E. N., and Beardmore, H. E., Supportive use of total parenteral alimentation in children with severe pancreatic injuries, *J. Pediatr. Surg.*, 11, 961, 1976.

Driscoll, R. H. and Rosenberg, I. H., Total parenteral nutrition in inflammatory bowel disease, *Med. Clin. N. Am.*, 62, 185, 1978.

Elwyn, D. H., Nutritional requirements of adult surgical patients, *Crit. Care Med.*, 8, 9, 1980.

Fischer, J. E., Foster, G. S., Abel, R. M., Abbott, W. M., and Ryan, J. A., Hyperalimentation as primary therapy for inflammatory bowel disease, *Am. J. Surg.*, 125, 165, 1973.

Fleming, C. R., McGill, D. B., and Berkner, S., Home parenteral nutrition as primary therapy in patients with extensive Crohn's disease of the small bowel and malnutrition, *Gastroenterology*, 73, 1077, 1977.

Force, R. A. and Shizgal, H. M., Assessment of malnutrition, *Surgery*, 88, 17, 1980.

Gamble, J. L., *Chemical Anatomy, Physiology and Pathology of Extracellular Fluid: A Lecture Syllabus*, 5th ed., Harvard University Press, Cambridge, Mass., 1947.

Gazzaniga, A. B., Bartlett, R. H., and Shobe, J. B., Nitrogen balance in patients receiving either fat or carbohydrate for total intravenous nutrition, *Ann. Surg.*, 182, 163, 1975.

Glynn, M. F. X., Langer, B., and Jeejeebhoy, K. N., Therapy for thrombotic occlusion of long-term intravenous alimentation catheters, *J. Parenteral Enteral Nutr.*, 4, 387, 1980.

Golden, M. H. N., Golden, B., Harland, P. S. E. G., and Jackson, A. A., Zinc and immunocompetence in protein-energy malnutrition, *Lancet*, 1, 1226, 1978.

Goodgame, J. T. and Fischer, J. E., Parenteral nutrition in the treatment of acute pancreatitis. Effect on complications and mortality, *Ann. Surg.*, 186, 651, 1977.

Greenberg, G.R. and Jeejeebhoy, K. N., Intravenous protein-sparing therapy in patients with gastrointestinal disease, *J. Parenteral Enteral Nutr.*, 3, 427, 1979.

Garrow, J. S., Fletcher, K., and Halliday, D., Body composition in severe infantile malnutrition, *J. Clin. Invest.*, 44, 417, 1965.

Graham, G. G., Cordano, A., Blizzard, R. M., and Cheek, D. B., Infantile malnutrition: changes in body composition during rehabilitation, *Pediatr. Res.*, 3, 579, 1969.

Greenberg, G. R., Marliss, E. B., Anderson, G. H., Langer, B., Spence, W., Tovee, E. B., and Jeejeebhoy, K. N., Protein-sparing therapy in the post-operative patient: effects of added hypocaloric glucose or lipid, *N. Engl. J. Med.*, 294, 1411, 1976.

Grischkan, D., Steiger, E., and Fazio, V., Maintenance of home hyperalimentation in patients with high-output jejunostomies. *Arch. Surg.*, 114, 838, 1979.

Heatley, R. V., Williams, R. H. P., and Lewis, M. H., Pre-operative intravenous feeding — a controlled trial, *Postgrad. Med. J.*, 55, 541, 1979.

Hill, G. L., King, R. F. G. J., Smith, R. C., Smith, A. H., Oxby, C. B., Sharafi, A., and Burkinshaw, L., Multi-element analysis of the living body by neutron activation analysis — application to critically ill patients receiving intravenous nutrition, *Br. J. Surg.*, 66, 868, 1979.

Holman, R. J., Essential fatty acid deficiency, in *Progress in Chemistry of Fats and Other Lipids,* Vol. 9 (Part 2), Holman, R. T., Ed., Pergamon Press, Oxford, 1968, 275.

Irvin, T. T., Effects of malnutrition and hyperalimentation on wound healing, *Surg. Gynecol. Obstet.*, 146, 33, 1978.

Izsak, E. M., Shike, M., Roulet, M., and Jeejeebhoy, K. N., Pancreatitis in association with hypercalcemia in patients receiving total parenteral nutrition, *Gastroenterology,* 79, 555, 1980.

James, J. C. and Hume, D. M., Anemia and neutropenia caused by copper deficiency, *Ann. Intern. Med.*, 80, 470, 1974.

Jamum, S., *Protein-Losing Gastroenteropathy,* Blackwell Scientific, Oxford, 1963.

Jeejeebhoy, K. N., Cause of hypoalbuminaemia in patients with gastrointestinal and cardiac disease, *Lancet,* 1, 343, 1962.

Jeejeebhoy, K. N., Bruce-Robertson, A., Ho, J., and Sodtke, U., The comparative effects of nutritional and hormonal factors on the synthesis of albumin, fibrinogen and transferrin, in *Protein Turnover,* Associated Sciences Publishers, Amsterdam, 1973, 217.

Jeejeebhoy, K. N., Zohrab, W. J., Langer, B., Phillips, M. J., Kuksis, A., and Anderson, G. H., Total parenteral nutrition at home for 23 months without complication and with good rehabilitation. A study of technical and metabolic features, *Gastroenterology,* 65, 811, 1973.

Jeejeebhoy, K. N., Langer, B., Tsallas, G., Chu, R.C., Kuksis, A., and Anderson, G. H., Total parenteral nutrition at home: studies in patients surviving 4 months to 5 years, *Gastroenterology,* 71, 943, 1976.

Jeejeebhoy, K. N., Anderson, G. H., Nakhooda, A. F., Greenberg, G. R., Sanderson, I., and Marliss, E. B., Metabolic studies in total parenteral nutrition with lipid in man: comparison with glucose, *J. Clin. Invest.*, 57, 125, 1976.

Kay, R. G. and Tasman-Jones, C., Acute zinc deficiency in man during intravenous alimentation, *Aust. N. Z. J. Surg.*, 45, 325, 2975.

Kelts, D. G., Grand, R. J., Shen, G., Watkins, J. B., Werlin, S. L., and Boehme, C., Nutritional basis or growth failure in children and adolescents with Crohn's disease, *Gastroenterology,* 76, 720, 1979.

Keys, A., Brozek, J., Hanschel, A., Michelson, O., and Taylor, H. L., *The Biology of Human Starvation,* University of Minnesota Press, Minneapolis, 1950.

Kinney, J. M., Long, C. L., and Duke, J. H., Carbohydrate and nitrogen metabolism after injury, in *Energy Metabolism in Trauma,* Knight, J., Ed., J. & A. Churchill, London, 1970, 103.

Kirsch, R. E., Saunders, S. J., Frith, L., Wicht, S., Kelman, L., and Brock, J. F., Plasma amino acid concentration and the regulation of albumin synthesis, *Am. J. Clin. Nutr.*, 22, 1559, 1969.

Kirschner, B. S., Voinchet, O., and Rosenberg, I., Growth retardation in inflammatory bowel disease, *Gastroenterology,* 75, 504, 1978.

Kishi, H., Nishii, S., Ono, T. et al., Thiamin and pyridoxine requirements during intravenous hyperalimentation, *Am. J. Clin. Nutr.*, 32, 332, 1979.

Layden, T., Rosenberg, J., Nemchausky, B., Elson, C., and Rosenberg, I. H., Reversal of growth arrest in adolescents with Crohn's disease after parenteral alimentation, *Gastroenterology,* 70, 1017, 1976.

Lewis, K. O., The nature of copper complexes in bile and their relationship to the absorption and excretion of copper in normal subjects and in Wilson's disease, *Gut,* 14, 221, 1973.

Long, J. M., Wilmore, D. W., Mason, A. D., Jr., and Pruitt, B. A., Jr., Fat-carbohydrate interaction: effects on nitrogen-sparing in total intravenous feeding, *Surg. Forum,* 25, 61, 1974.

Lowry, S. F., Goodgame, J. T., Maher, M. M. et al., Parenteral vitamin requirements during intravenous feeding, *Am. J. Clin. Nutr.*, 31, 2149, 1978.

McFie, J. and Hill, G. L., Glucose or fat as a non-protein energy source? A controlled clinical trial in gastroenterological patients requiring intravenous nutrition, *Gastroenterology,* in press.

McLoughlin, G. A., Wu, A. V., Saporoschetz, I., Nimberg, R., and Mannick, J. A., Correlation between anergy and a circulating immunosuppressive factor following major surgical trauma, *Ann. Surg.*, 190, 297, 1979.

McNeill, K. G., Mernagh, J. R., Jeejeebhoy, K. N., Wolman, S. L., and Harrison, J. E., In vivo measurements of body protein based on the determination of nitrogen by prompt gamma analysis, *Am. J. Clin. Nutr.*, 32, 1955, 1979.

McNeill, K. G., Harrison, J. E., Mernagh, J., Stewart, S., and Jeejeebhoy, K. N., Changes in body protein, body potassium and lean body mass during total parenteral nutrition (TPN), *J. Parenteral Enteral Nutr.*, 1, 1981.

Meakins, J. L., Christou, N. V., Shizgal, H. M., and MacLean, L. D., Therapeutic approaches to anergy in surgical patients, *Ann. Surg.*, 190, 286, 1979.

Mertz, W. and Roginski, E. E., Chromium metabolism: the glucose tolerance factor, in *Newer Trace Elements in Nutrition,* Mertz, W. and Cornatzer, W. E., Eds., Marcel Dekker, New York, 1971, 123.

Messing, B., Latrive, J. P., Bitoun, A., Galian, A., and Bernier, J. J., La steatose hepatique au course de la nutrition parenterale totale depend-elle de l'apport calorique lipidique? *Gastroenterol. Clin. Biol.*, 3, 719, 1979.

Messing, B., Bitoun, A., Galian, A., Mary, J. Y., Goll, A., and Bernier, J. J., La steatose hepatique au cours de la nutrition parenteral depend-elle de l'apport calorique glucidique? *Gastroenterol. Clin. Biol.*, 1, 1015, 1977.

Miller, C. L., Immunological assays as measurements of nutritional status: a review, *J. Parenteral Enteral Nutr.*, 2, 554, 2978.

Moore, F. D., Energy and the maintenance of the body cell mass, *J. Parenteral Enteral Nutr.*, 4, 228, 1980.

Moss, G., Bierenbaum, A., Bova, F., and Slavin, J. A., Post-operative metabolic patterns following immediate total nutritional support: hormone levels, DNA synthesis, nitrogen balance and accelerated wound healing, *J. Surg. Res.*, 21, 383, 1976.

Munro, H. N., General aspects of the regulation of protein metabolism by diet and by hormones, in *Mammalian Protein Metabolism*, Vol. 1, Munro, H. N. and Allison, J. B., Eds., Academic Press, New York, 1964, 381.

Nichoalds, G. E., Meng, H. C., and Caldwell, M. D., Vitamin requirements in patients receiving total parenteral nutrition, *Arch. Surg.*, 112, 1061, 1977.

Patel, D., Anderson, G. H., and Jeejeebhoy, K. N., Amino acid adequacy of parenteral casein hydrolysate and oral cottage cheese in patients with gastrointestinal disease as measured by nitrogen balance and blood aminogram, *Gastroenterology*, 65, 427, 1973.

Reilly, J., Ryan, J. A., Strobe, W. et al., Hyperalimentation in inflammatory bowel disease, *Am. J. Surg.*, 131, 192, 1976.

Richardson, D. P., Scrimshaw, N. S., and Yeung, V. R., The effect of dietary sucrose on protein utilization in healthy young men, *Am. J. Clin. Nutr.*, 33, 264, 1980.

Roulet, M., Shike, M., Marliss, E. B., Todd, T. R. J., Mahon, W. A., Anderson, G. H., Stewart, S., and Jeejeebhoy, K. N., Energy and protein metabolism in critically ill, malnourished and septic patients, *J. Parenteral Enteral Nutr.*, 1, 1981.

Rudman, D., Millikan, W. J., Richardson, T. J., Bixler, T. J., Stackhouse, W. J., and McGarrity, W. C., Elemental balances during intravenous hyperalimentation of underweight adult subjects, *J. Clin. Invest.*, 55, 94, 1975.

Schoenheimer, R., in *The Dynamic State of Body Constituents*, Harvard University Press, Cambridge, Mass., 1946.

Shike, M., Harrison, J. E., Sturtridge, W. C., Tam, C. S., Bobechko, P. E., Jones, G., Murray, T. M., and Jeejeebhoy, K. N., Metabolic bone disease in patients receiving long-term total parenteral nutrition, *Ann. Int. Med.*, 92, 343, 1980.

Shizgal, H. M., Spannier, A. H., Hanes, J., and Wood, C. D., Indirect measurement of total exchangeable potassium, *Am. J. Physiol.*, 233, F253, 1977.

Silvas, S. E. and Parpgas, P. D., Paresthesias, weakness, seizures and hypophosphatemia in patients receiving hyperalimentation, *Gastroenterology*, 62, 513, 1972.

Soeters, P. B., Parenteral nutrition and gastrointestinal fistulas, in *Current Concepts in Parenteral Nutrition*, Greep, J. M., Soeters, P. B., Wesdorp, R. I. C., Phaf, C. W. R., and Fischer, J. E., Eds., Martinus Nijhoff, The Hague, Netherlands, 1977, 99.

Soeters, P. N., Ebeid, A. M., and Fischer, J. E., Review of 404 patients with gastrointestinal fistulae. Impact of parenteral nutrition, *Ann. Surg.*, 190, 189, 1979.

Solomans, N. W., Layden, T. J., Rosenberg, I. H., Vokhactu, K., and Sandstead, H. H., Plasma trace metals during total parenteral alimentation, *Gastroenterology*, 70, 1022, 1976.

Spark, R. F., Arky, R. A., Boulter, P. R., Savdek, C. D., and O'Brian, J. T., Renin, aldosterone and glucagon in the natriuresis of fasting, *N. Engl. J. Med.*, 292, 1335, 1975.

Vanamee, P., Shils, M. E., and Burke, A. W., Multivitamin preparation for parenteral use. A statement by the nutrition advisory group, *J. Parenteral Enteral Nutr.*, 3, 258. 1979.

van Rij, A. M., Thomson, C. D., McKenzie, J. M., and Robinson, M. F., Selenium deficiency in total parenteral nutrition, *Am. J. Clin. Nutr.*, 32, 2076, 1979.

van Way, C. W., III, Buerk, C. A., Peterson, R., and Dresler, C., Nitrogen balance and electrolyte requirements in intralipid-based hyperalimentation, *J. Parenteral Enteral Nutr.*, 3, 174, 1979.

Veverbrants, E. and Arky, R. A., Effects of fasting and refeeding. I. Studies on sodium, potassium water excretion on a constant electrolyte and fluid intake, *J. Clin. Endocrinol.*, 29, 55, 1969.

Vogel, C. M., Corwin, T. R., and Bane, A. E., Intravenous hyperalimentation in the treatment of inflammatory diseases of the bowel, *Arch. Surg.*, 108, 460, 1974.

Voit, C., in *Hermann's Handbuch der Physiologie Leipzic*, 6, 117, 1881.

Waldmann, J. A., in *Gastroenterology*, Vol. 2, 3rd ed., Bockus, H. L., Ed., W. B. Saunders, Philadelphia, 1976, 361.

Wannemacher, R. W., Jr., Kaminski, M. V., Jr., Dinterman, R. W., Bostian, K. A., and Hadick, C. L., Protein-sparing therapy during pneumococcal infection in rhesus monkeys, *J. Parenteral Enteral Nutr.*, 2, 507, 1978.

Wolfe, R. R., Durkot, M. J., Allsop, J. R., and Burke, J. F., Glucose metabolism in severely burned patients, *Metabolism*, 28, 1031, 1979.

Wolman, S. L., Anderson, G. H., Marliss, E. B., and Jeejeebhoy, K. N., Zinc in TPN: requirements and metabolic effect, *Gastroenterology*, 76, 458, 1979.

Woolfson, A. M. J., Heatley, R. V., and Allison, S. P., Insulin to inhibit protein catabolism after injury, *N. Engl. J. Med.*, 300, 14, 1979.

Yeung, C. K., Smith, R. C., and Hill, G. L., Effect of an elemental diet on body composition: a comparison with intravenous nutrition, *Gastroenterology*, 77, 652, 1979.

Zohrab, W. J., McHattie, J. D., and Jeejeebhoy, K. N., Total parenteral alimentation, with lipid, *Gastroenterology*, 64, 583, 1973.

Chapter 5

PRACTICAL ASPECTS OF ALIMENTATION CATHETERS

Zane Cohen and Bernard Langer

TABLE OF CONTENTS

I. INTRODUCTION

During the past decade, ever-increasing value has been placed on the establishment of central venous catheters. The major indications for a central venous line include: its use in monitoring the critically ill patient, the rapid administration of fluids, parenteral alimentation, extensively burned patients with no peripheral access, obesity, the patient with extremely small peripheral veins, and infusion of hyperosmolar solutions. We will concern ourselves, in this chapter, only with the use of central catheters for the purpose of parenteral nutrition.

II. ROUTES OF ADMINISTRATION

A. Temporary

Peripheral: Nutritional support can be given by cannulation of a peripheral vein. Indications include those patients requiring short-term parenteral nutrition of less than 1 week in duration and those patients who present a high risk for central venous catheterization or have no central venous access. However, the peripheral route has limited use. The administration of hypertonic solutions results in pain and sclerosis of the vein, necessitating frequent changing of i.v. sites. Therefore, solutions are used with decreased tonicity — a 6% amino acid and 12% dextrose solution is used as compared with the centrally administered 4.2% amino acid and 25% dextrose. This alone limits the total number of calories that can be supplied by utilizing this route of administration.

Central: In order to supply sufficient calories to achieve growth, solutions with high osmolarities are required. The concentrated solutions are not suitable for long-term administration through short catheters in peripheral veins. Long catheters threaded through antecubital veins into the SVC cause axillary vein sclerosis and thrombophlebitis. Therefore, long-term supply of intravenous alimentation was not practical until techniques of central venous infusion were developed.

Since the original description by Aubaniac in 1952 of the technique of infraclavicular subclavian vein insertion, numerous reports have appeared in the literature. Dudrick adapted subclavian vein catheterization in 1968 as a method for long-term parenteral nutrition. The tip of the catheter is positioned in the superior vena cava so that hyperosmolar solutions can be infused quickly and diluted rapidly. The routes of insertion of central venous catheters for administration of TPN are as follows.

Subclavian Vein — Infraclavicular Approach: This is the author's preferred approach. The details of catheter insertion are presented below.

Subclavian Vein — Supraclavicular Approach: This technique was originally described by Yoffa in 1965. The advantages of this particular approach are that the complication of pneumothorax is less frequent and that the risk of malposition of the catheter tip is also lessened. Although reported complications seem to be less with this approach, it has the disadvantage that the entry site of the catheter is in the hollow of the supraclavicular fossa making access for dressing changes difficult. Although this can be corrected by tunneling the catheter over the clavicle on the anterior chest wall, the author prefers the infraclavicular approach. When experienced personnel carry out infraclavicular subclavian vein puncture the complications of insertion should be minimal, offering no real advantage for the supraclavicular approach.

Internal Jugular Vein Catheterization: This approach does obviate the complications of pneumothorax, hemothorax, and brachial nerve injury. The most frequent complication has been arterial puncture. The disadvantage is again related to the entry site of the catheter and the resulting inhibition to neck movements created by this entry site and its overlying dressings. Details of catheter insertion using this site have been reported by many authors.

B. Permanent

A system for long-term total parenteral nutrition has the potential for maintaining excellent health in patients who are unable to take nourishment orally. The advantage of long-term permanent ambulatory alimentation has broadened our scope for those diseases heretofore thought untreatable. Solutions have become available to provide sufficient carbohydrate, fat, protein, and trace metals to allow maintenance of health. Access to the circulation for long-term alimentation was first described by Rhoads in 1969. In 1970, Scribner described the concept of the "artificial gut" using arteriovenous (A-V) shunts as a mode of access to the circulation. However, the problems associated with A-V shunts described by Scribner and others has forced the development of floating superior vena cava catheters for long-term vascular access. The difficulties encountered with A-V shunts are related mainly to clotting of the shunt. Furthermore, many patients with chronic intestinal diseases have badly damaged or destroyed veins in the arms from previous infusions.

The Toronto General Hospital technique for placement of permanent alimentation catheters is described below. The development of a "home alimentation" program has been carried out in relatively few specialized units throughout the world. At the Toronto General Hospital, technical developments related to the infusion apparatus and the mode of vascular access have made long-term ambulatory alimentation possible. Patients must be suitably chosen and properly educated in the management of their apparatus. The infusions are performed by the patient at home without professional supervision. When successful, the patient can be fully rehabilitated and again successfully integrated into a normal active life, achieving independence, comfort, and a return to physical and mental health.

Vascular access has been achieved by using Silastic® "Langer" catheters usually inserted through the cephalic vein and into the superior vena cava just proximal to the right atrium. At times, a branch of the internal jugular vein has been used as the insertion site. If there is no vascular access through the cephalic or internal jugular veins, the inferior vena cava can be cannulated either directly or through a branch of the external iliac vein.

III. PLACEMENT OF CATHETERS

A. Temporary

The author prefers to deliver parenteral nutrition through a central venous catheter inserted into the superior vena cava via the infraclavicular approach. This is a comfortable line, easily managed by the patient and the "alimentation team". It in no way inhibits neck movements.

The method of insertion of this catheter is outlined below.

1. Preparation

The placement of a central venous catheter for parenteral nutrition is usually not an emergency procedure. The patient should therefore be well hydrated in order to provide adequate venous pressure to fill the vein. The procedure is carried out at the bedside. The foot of the bed is elevated to 15° so that the Trendelenburg position is achieved. This will also aid in the filling of the subclavian veins. The head is turned to the side opposite to that of the catheter insertion. The hands and arms are kept at the sides. The skin over the upper chest is shaved and carefully prepared with povidone-iodine solution. The area is draped appropriately. The physician is then gloved, gowned, and masked. All other personnel within the room, including the patient, are masked.

2. Landmarks

The following are identified: the suprasternal notch, the lower anterior border of the sternomastoid muscle, and the junction between the medial and middle thirds of the clavicle. The subclavian vein in the adult is approximately 3 to 4 cm long and is directly posterior to the medial third of the clavicle. In this position it is separated from the subclavian artery by the scalenus anterior muscle. Directly posterior to the subclavian artery are the brachial plexus, first rib, and apex of lung. More medially, the apical pleura is in direct contact with the postero-inferior aspect of the subclavian vein. It is thus important that the needle and catheter puncture the vein directly posterior to the clavicle using the junction of the medial and middle thirds of the clavicle as an external landmark.

3. Identifying the Vein

Using 1% Xylocaine® local anesthetic, a skin wheal is raised 1 to 1.5 cm below the junction of the medial and middle thirds of the clavicle. A number 22 needle attached to a 2-cc syringe is then used to further infiltrate local anesthetic into the periosteum over the postero-inferior aspect of the clavicle. The needle should be directed toward, and a centimeter above, the sternal notch (Figure 1). A good point of reference can be established by firmly pressing the fingertip into the suprasternal notch to locate the deep side of the superior angle of the clavicle and directing the course of the needle just above this point (Figure 2). It is essential that the postero-inferior aspect of the clavicle be identified before piercing the vein. The needle should be inserted at less than 10° from the horizontal position to mininize risk of injury to underlying structures. Once the vein is located, the angle of insertion is noted.

4. Venipuncture

At the present time our preference is to use a polyvinyl Intracath® (16 gauge) with a 14-gauge needle. A preliminary measurement on the chest wall over the course of the veins is helpful in estimating where the tip of the catheter should be positioned. The needle is removed from the Intracath® and attached to a 2-cc syringe. While maintaining a slight negative pressure by withdrawing the plunger of the syringe, the needle is again advanced in a similar manner and direction to that described above. The needle must always be directed in a straight line. It should never be moved from side to side as this might cause laceration of the vein by the beveled tip. When the vein is entered, blood will flow freely back into the syringe.

5. Insertion of Line

The needle is grasped with an instrument to prevent displacement. The patient is asked to exhale and hold his/her breath, at which time the syringe is disconnected from the needle. A finger is placed over the needle opening to prevent bleeding or air embolism. The catheter is inserted through the needle and advanced into the vein. There is no voluntary control over the direction taken by the advancing catheter. There are however specific principles which should be emphasized at this stage:

1. Gentle pressure should be applied to the catheter as it is inserted. If it does not advance easily, do not force it.
2. If the catheter does not advance easily, always remove the catheter and the needle together. The catheter should never be withdrawn through the needle, since its end may be sheared off accidentally by the needle resulting in a catheter embolus.

The catheter is advanced into what is thought to be the correct position. The catheter is held in this position and the needle is withdrawn over it. The protective guard is applied over the tip of the needle to prevent shearing of the catheter. The stylet is removed and a syringe is attached to the catheter. Examination of the stylet for a gentle curve will indicate the correct direction of the catheter. Blood is once again aspirated into the syringe to confirm that the catheter is located within the vein (Figure

3). An intravenous solution is then infused through an infusion set attached to the catheter. Any isotonic solution can be infused at this stage. The intravenous solution bag or bottle is lowered below that of the bed. Retrograde flow of blood ensures that the catheter is located within a central vein. The protective guard and the catheter are sutured in place (Figure 4). All connections are secured with tape. An occlusive dressing (Op-Site®) is placed over the area (Figure 5).

A chest X-ray is then obtained to confirm the location of the tip of the catheter within the superior vena cava and to ensure that a pneumothorax is not present. If at any time air is aspirated during the venipuncture, or if the patient develops chest pain following catheter insertion, one should immediately obtain a chest X-ray. In the event of a pneumothorax, emergency insertion of a chest tube may be necessary.

B. Permanent

It is now generally agreed that a silicone rubber (Silastic®) catheter placed in the superior or inferior vena cava via the subclavian, jugular, or external iliac veins constitutes the most satisfactory vascular access system. Most commonly the cephalic vein in the deltopectoral region has been used as access to the circulation. The method of insertion and placement of the permanent alimentation catheter is outlined below.

1. Preparation

The procedure is carried out in the operating room under general anesthesia. An image intensifier is required in the operating room for correct placement of the catheter. The evening before surgery, the alimentation nurse will mark on the anterior chest wall the most suitable exit site for the catheter. The site chosen depends on the body habitus and the desire of the patient to conceal the catheter. The entire chest and both shoulders are shaved and washed twice with Betadine®. No food or drink is allowed after midnight. Antibiotics may be used 1 hr preoperatively and for 48 hr postoperatively.

In the operating room, the patient is positioned on a table which allows room for the use of the image intensifier. Following anesthesia, a towel is rolled and placed longitudinally between the shoulder blades. This allows for easier access to the cephalic vein. The patient is anesthetized and the entire chest and both shoulder regions are suitably prepared and draped. A curvilinear incision approximately 3 cm in length is made in the deltopectoral groove. Dissection is carried down between the deltoid and pectoral muscles. The cephalic vein is usually quite easily identified (Figure 6). On occasion it is necessary to incise the clavicular fibers of the pectoralis muscle in order to expose a suitable length of vein for cannulation. A small incision is made over the designated exit site (Figure 7) and the catheter is tunneled subcutaneously to the deltopectoral region (Figure 8). The catheter utilized (Langer catheter) is a Silastic® tube 100 cm in length, with an internal diameter of 0.76 mm and an external diameter of 2.72 mm. It is connected to a permanent plastic Luer-lock hub (available through Extracorporeal Medical Specialties Inc.). However, the same principle can be used with other silicone rubber catheters, such as Broviac and Hickman catheters. It is important to wash the powder from the operating surgeon's gloves prior to catheter manipulation. Once the catheter appears in the deltopectoral region, an approximation is made of the length of catheter required for correct positioning in the superior vena cava. The cephalic vein is then ligated distally and mobilized proximally. A 3-0 ligature is passed around the mobilized portion of the vein. This ligature is looped once, reflected proximally, and tagged. The catheter is then flushed with a heparinized saline solution. The anterior wall of the cephalic vein is grasped and a small transverse incision is made. The catheter is inserted to a predetermined length. Aspiration of the blood ensures that the catheter is within the vascular system. The catheter is once again flushed

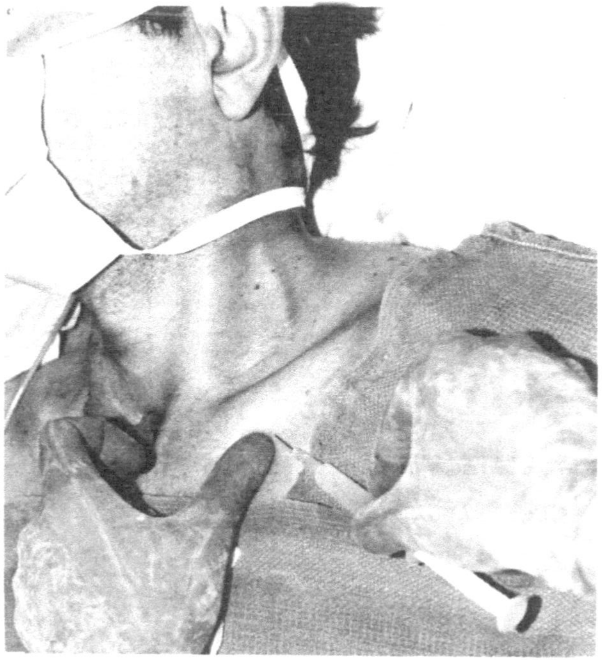

FIGURE 1

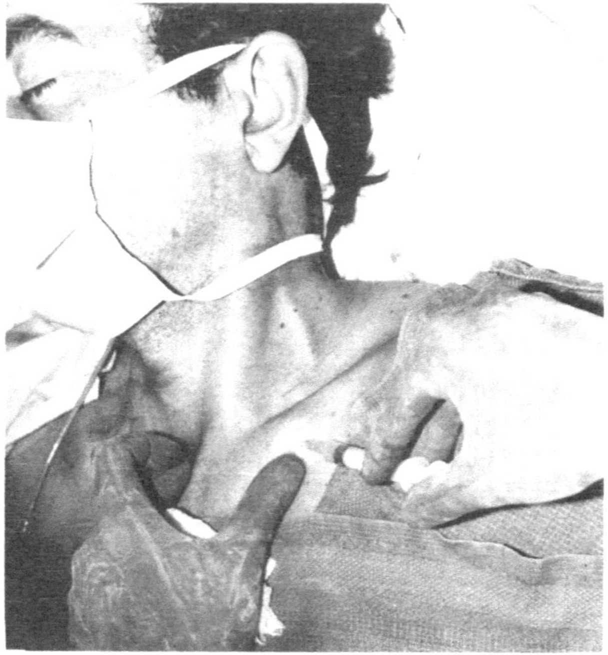

FIGURE 2

FIGURE 1. A 2-cc syringe attached to a 22-gauge needle is used to infiltrate local anesthetic 1.0 to 1.5 cm below the junction of the medial and middle thirds of the clavicle. FIGURE 2. To identify the vein, the needle is shown to be directed toward, and a centimeter above, the sternal notch. The needle should be inserted at less than 10° from the horizontal position.

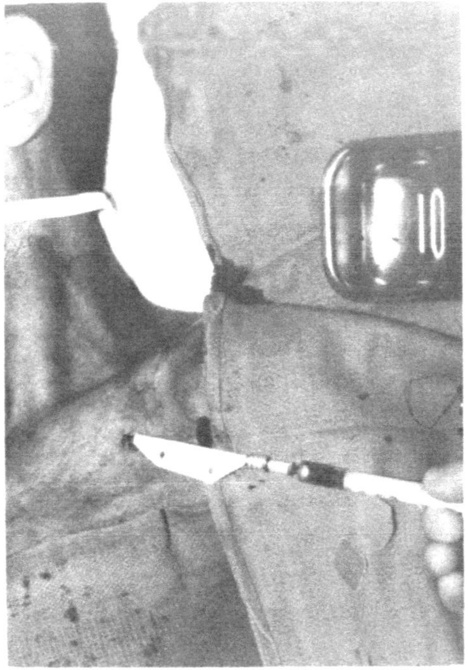

FIGURE 3

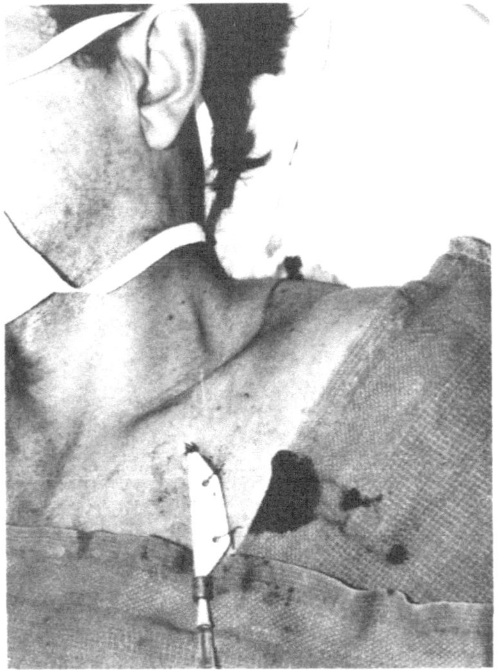

FIGURE 4

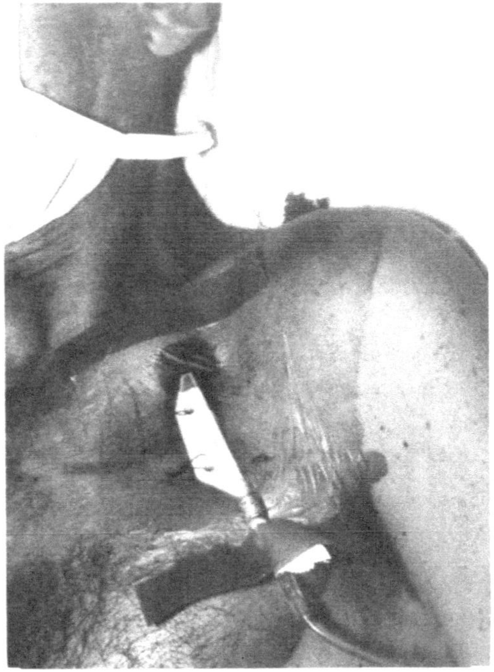

FIGURE 5

FIGURE 3. After placement of the catheter and withdrawal of the needle, the protective guard has been applied. Blood is once again aspirated into the syringe to confirm its location within the vein. FIGURE 4. The protective guard is shown sutured in position and an infusion set has been connected to the catheter. FIGURE 5. The connection of the infusion set to the catheter has been secured with tape and an occlusive dressing has been applied.

with heparinized saline and secured loosely to the vein with the proximal ligature (Figure 9). The catheter is injected with radiopaque dye (Hypaque®) and positioned correctly with the aid of the image intensifier so that it lies well down in the superior vena cava just above the right atrium. It is important to emphasize that the catheter should neither be positioned high in the superior vena cava nor in the right atrium. Caval thrombosis or incorporation into the endocardium can result from catheter malposition. The proximal ligature is further secured around the catheter and vein. The catheter is also secured to the pectoralis major muscle. It is important that ligatures around the catheter do not occlude its lumen. The subcutaneous tissue and skin are closed in a conventional manner. Occasionally an additional small incision is made midway between the exit site and the deltopectoral region. The catheter is further secured to the surrounding tissue at this site to prevent catheter migration and subsequent infection.

A small Teflon® patch is sutured to the skin approximately 1 cm below the exit site. The catheter is secured to the Teflon® patch with two 1-0 silk sutures and Silastic® medical adhesive is placed over the catheter and silk ligatures (Figure 10). This maneuver is carried out to prevent movement of the catheter at the exit site. The catheter is connected to an infusion set and suitable dressings are applied. All junctions are reinforced with tape.

Postoperatively, the position of the catheter is once again documented with a chest X-ray. Following this, the alimentation solutions may be infused.

IV. COMPLICATIONS

The number of complications related to parenteral nutrition is significant. Many published reports have documented these complications, the most frequent of which are catheter related. The following discussion will concentrate briefly on complications related to catheter insertion, catheter-related sepsis, and late complications of indwelling catheters.

A. Complications of Catheter Insertion

Injury to lung and pleura: Pneumothorax is the most common complication of cathether insertion. On occasion a tension pneumothorax can develop. Usually a pneumothorax will be diagnosed by the chest X-ray taken following catheter insertion. If small and causing no clinical symptoms, no treatment is required. The size of the pneumothorax can be documented by daily chest X-rays. If it enlarges, needle aspiration or insertion of a thoracostomy tube would be indicated. On the other hand, if a patient presents with dyspnea, chest pain, or cyanosis following catheter insertion, a major complication of the pleural space should be suspected. If warranted, insertion of a large bore needle or a thoracostomy tube should be performed to relieve a probable pneumothorax, tension pneumothorax, or hemothorax. If the catheter is in the correct position within the superior vena cava and if the pneumothorax is simply treated, there is no necessity for catheter removal.

Injury to the artery and vein: Injury to the subclavian artery with resultant external bleeding can be treated by removing the needle and/or catheter and applying direct pressure. The radial and ulnar pulses on the injured side should be checked frequently. Laceration of a major vein should be controlled by direct pressure. If the tip of the catheter lacerates the vein and terminates in the pleural space, infusion of fluid will quickly produce chest pain, dyspnea, and possibly shock. A chest X-ray and aspiration of the chest will confirm this diagnosis. This complication can be avoided if one lowers the intravenous infusion bottle or bag following catheter insertion, to observe reflux of blood. Treatment should include immediately discontinuing the infusion, removing the catheter, and monitoring the pleural space by a thoracostomy tube if necessary.

Injury to brachial plexus: Nerve damage reflected by radial, ulnar, or median nerve signs can be treated simply by removing the catheter.

Injury to thoracic duct: Injury to the thoracic duct from a left subclavian catheter will be obvious when clear lymph drains from the insertion site. Treatment involves removal of the catheter. Chylothorax has been reported and should be treated by evacuation and aspiration of the pleural space until lymphatic drainage ceases.

Injury to mediastinum: Insertion of a temporary catheter rarely may result in the production of a mediastinal hematoma. If large, this could cause compression of the superior vena cava. In this situation, emergency surgery is required to evaluate the hematoma and relieve the obstruction.

Embolus: Air embolism can occur at the time of catheter insertion when the syringe is removed prior to threading of the catheter. It can be avoided by having the patient perform a Valsalva maneuver during this step in the technique. Catheter embolism has occurred on two occasions in the author's experience. It is due to pulling back on the *catheter and adjusting its position while the needle is still in the vein. The catheter is* sheared off against the beveled tip of the needle and lodges in the vein. The treatment of choice for a catheter embolus is transvenous removal with a guidewire snare technique performed under image intensification in the cardiac catheterization unit.

Malposition of the catheter: The catheter must be positioned low down in the SVC. On occasion, during insertion the catheter will enter the ipsilateral internal jugular vein or cross over to the contralateral innominate vein. In either location infusion of hypertonic solutions can cause thrombophlebitis. If malposition occurs it should be corrected by repositioning under fluoroscopic control. If this is not possible the catheter should be removed and a new catheter reinserted on the contralateral side.

Cardiac complications: Cardiac arrythmias may occur if the catheter is positioned within the right atrium or ventricle. This complication can be avoided by correct positioning of the catheter within the SVC.

B. Septic Complications

Catheter sepsis can be defined as an episode of clinical sepsis in a patient receiving short- or long-term intravenous alimentation which resolved following stoppage of alimentation solutions, urokinase infusion of the catheter (see below), or removal of the catheter. Confirmatory evidence includes either a positive blood culture or positive cultures of removed catheter tips. In the author's experience, this complication results in the most patient morbidity. It is seen in the use of polyvinyl catheters for temporary alimentation and in Silastic® catheters used for permanent alimentation. The incidence of catheter sepsis varies from 2% to 7%.

The most common organisms involved in catheter sepsis are *Staphylococcus aureus* and *S. epidermidis*. Less-frequent organisms cultured are coliform, enteric, and fungi, particularly candida. Factors which may contribute to catheter sepsis include poor technique in catheter insertion, lack of adherence to postinsertion protocol, duration of catheterization, and patient population. Certain patients are at greater risk for catheter sepsis than others. High-risk patients include those receiving steroids, those with acute pancreatitis, anergic patients, and those with severe inflammatory bowel disease.

The most common portals of infection are those around the entrance site of the catheter. In patients receiving home alimentation sepsis has quite commonly been related to movement and migration of the catheter itself. In order to prevent migration, the catheter is routinely secured to the subcutaneous tissue between the catheter exit site and its insertion site into the venous system (see above). This maneuver has markedly reduced the incidence of migration and sepsis.

A low-grade fever in the absence of any other fever source should alert the physician to the possibility of catheter sepsis. A fever spike to 102° or above might also occur.

FIGURE 6

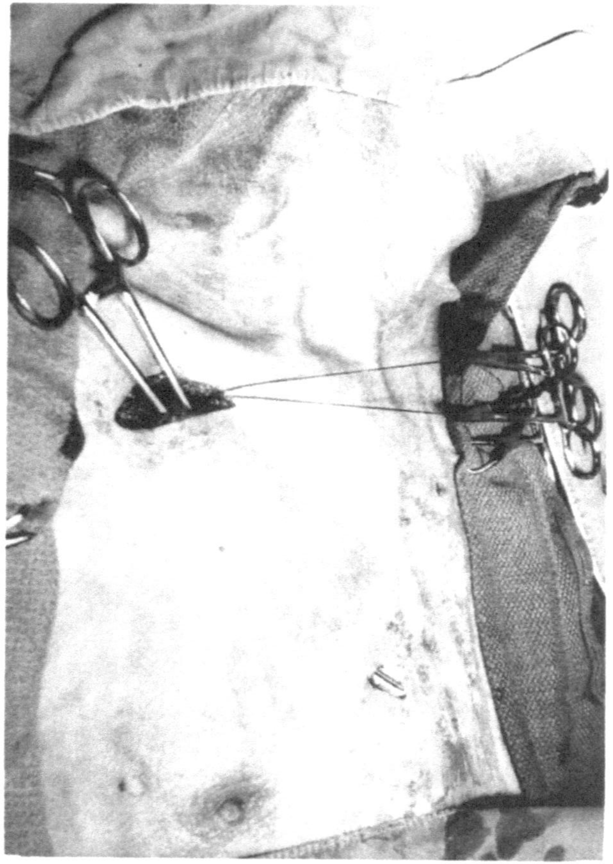

FIGURE 7

FIGURE 6. The cephalic vein is shown medial to the deltoid muscle. FIGURE 7. A small incision has been made over the proposed exit site for the catheter. The two incisions communicate via a subcutaneous tunnel.

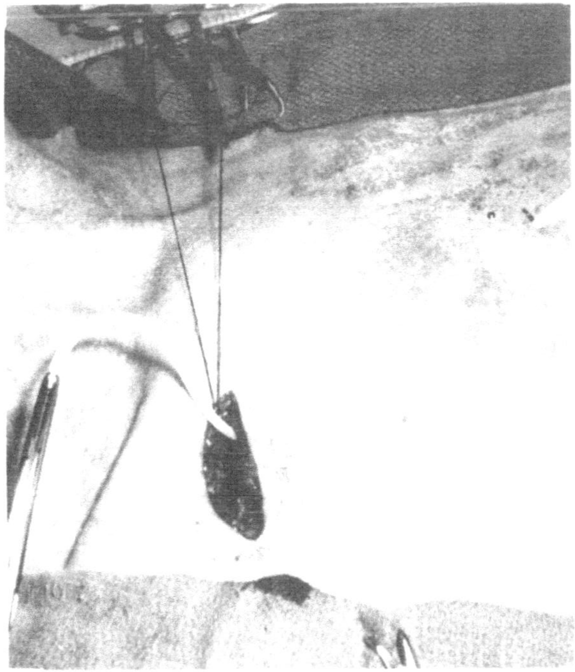

FIGURE 8

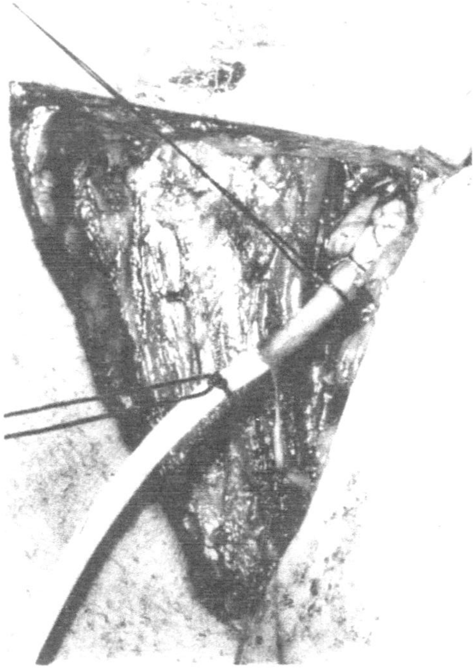

FIGURE 9

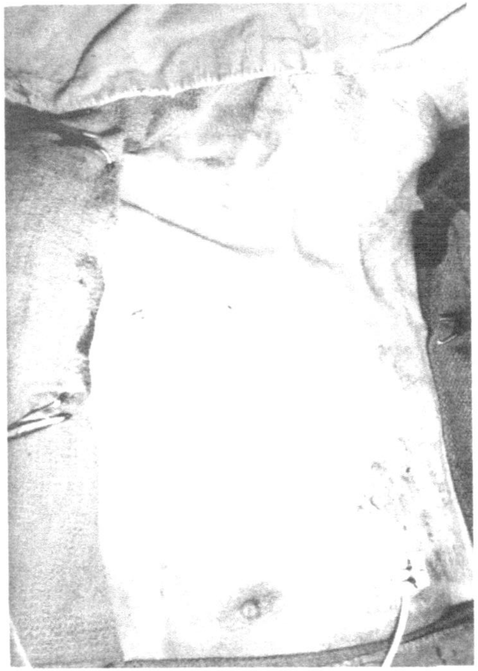

FIGURE 10

FIGURE 8. The Silastic® "Langer" catheter has been tunneled subcutaneously to the deltopectoral region. FIGURE 9. The catheter is shown inserted through the cephalic vein to the predetermined length (ligature on catheter). FIGURE 10. The catheter has been secured just below its exit site to a Teflon® patch, to prevent movement.

If catheter sepsis is suspected and other sources of fever have been ruled out, the parenteral alimentation should be discontinued and the infusate and tubing sent for culture. Maintenance intravenous fluids should be substituted. Peripheral blood cultures and retrograde cultures through the catheter are obtained. The exit site of the catheter is swabbed and cultured. Routine cultures of urine, sputum, mouth, and drainage sites are also obtained. If the blood cultures are positive or if the fever continues, the temporary subclavian catheter should be removed and reinserted after subsidence of the fever for 48 hr. If the blood cultures are negative and the temperature returns to normal, the Toronto General Hospital policy has been to resume total parenteral nutrition through the same line after 48 hr. In sepsis related to a permanent Silastic® line an association with thrombosis occurring at the catheter tip has been found. In this situation an infusion of urokinase is attempted (see below) to clear the catheter before deciding on its removal.

On occasion, removal of the catheter will effectively control sepsis. However, in certain circumstances fever and positive blood cultures will persist. In this situation, antibiotics are used. The catheter should not be reinserted for at least 48 hr after the temperature has returned to normal levels. Patients with established and persistent fungemia should be treated appropriately with intravenous Amphotericin B® over a 4 to 6 week period depending on the total dose given.

C. Late Complications

1. Venous and Catheter Thrombosis

Thrombosis of the SVC and its main tributaries can occur following long-term parenteral alimentation. The usual cause is malposition of the catheter and resultant phlebothrombosis and thrombophlebitis. This complication usually requires removal of the catheter. Anticoagulants are instituted to enhance clot lysis and prevent propagation. Recanalization normally occurs.

Thrombi can also occur at the tip of the catheter. Catheter thrombosis and sepsis are closely related complications of indwelling Silastic® catheters. Glynn et al. at the Toronto General Hospital described a procedure by which thrombi are dispersed with urokinase while the permanent Silastic® catheter remains *in situ*. Urokinase was successful in clearing the catheter in all 20 patients treated. The concomitant use of antibiotics cleared the sepsis. The procedure resulted in no detectable fibrinolysis or other complications.

2. Air Embolism

This can occur as a late complication if the intravenous line inadvertently becomes detached from the catheter or if the catheter itself is inadvertently removed. The patient is usually found collapsed with the intravenous line detached from the catheter. The physician or nurse should immediately cover the catheter and place the patient in the Trendelenburg position with the right side up. It is imperative to place the patient in the left lateral decubitus position to prevent air obstructing the pulmonary artery. If a cardiac arrest has occurred and the patient does not respond quickly, then aspiration of the air from the right ventricle is indicated.

VI. HOME PARENTERAL NUTRITION PROGRAM

Total parenteral alimentation has been one of the great advances during the past decade. It has enabled us to provide nutrition and growth for those who cannot feed enterally. The advances made through the use of permanent home parenteral nutrition have enabled patients, thought previously to be untreatable, to live nearly normal, active lives.

From 1970 to 1980, 47 patients have been entered into the Home Parenteral Nutrition (HPN) program at the Toronto General Hospital. They receive all or most of their nutrition via an indwelling venous catheter. There are 22 males and 25 females in the program, ranging in age from 17 to 76 years (mean 41.4 years). Indications for HPN included the short gut syndrome with resultant malabsorption in 35 patients, and chronic bowel obstruction in 12 patients.

A total of 80 catheters have been placed in the 47 patients and 42 catheters have been removed during the 10-year period. Of these, 10 have been removed because of remission of disease, 15 have been removed due to catheter sepsis, 6 because of catheter migration, 5 for venous thrombosis, and 3 for catheter blockage. Currently 27 patients have functioning catheters. A total of 11 patients have died while receiving HPN from causes unrelated to the parenteral nutrition system.

In a total of 88 patient-years, 96.6% of patients' time has been spent outside of the hospital. Of the 47 patients, 63% have returned to previous employment, 6% retired, and 30% remained disabled by their disease.

The first decade of experience has shown HPN to be an acceptable form of nutritional support in selected patients and has succeeded in returning many to their previous lifestyle out of the hospital.

READING LIST

Abbott, W. M., Ryan, J. A., Jr., Sollassol, C., and Joyeux, H., *Total Parenteral Nutrition*, Fischer, J. E., Ed., Little, Brown, Boston, 1976.

Abel, R. M., Beck, C. H., Abbott, W. M. et al., Improved survival from acute renal failure following treatment with intravenous essential L-amino acids and glucose, *N. Engl. J. Med.*, 288, 695, 1973.

Aubaniac, R., L'injection intraveineuse sous-claviculaire, *Presse Med.*, 60, 1456, 1952.

Brinkman, A. J. and Costley, D. O., Internal jugular venipuncture, *JAMA*, 223(2), 182, 1973.

Broviac, J. W., Cole, J. J., and Scribner, B. H., A silicone rubber atrial catheter for prolonged parenteral alimentation, *Surg. Gynecol. Obstet.*, 136, 602, 1973.

Broviac, J. W. and Scribner, B. H., The artificial gut: three years experience with total parenteral nutrition in the home, Int. Cong. Parenteral Nutrition, Montpellier, France, September, 1974.

Dudrick, S. J., Copeland, E. M., and MacFadyen, B. V. J., The nutritional care of the cancer patient, in *Current Concepts in Parenteral Nutrition*, Greep, J. M., et al., Eds., Marinus Nijoff, The Hague, 1977, 187.

Dudrick, S. J. and Ruberg, R. L., Principles and practice of parenteral nutrition, *Gastroenterology*, 61(6), 901, 1971.

Dudrick, S. J., Wilmore, D. W., Vars, H. M., and Rhoads, J., Can intravenous feeding as the sole means of nutrition support growth in the child and restore weight loss in an adult? *Ann. Surg.*, 6, 974, 1969.

Fischer, J. E., Nutritional support in the seriously ill patient, *Curr. Probl. Surg.*, 17(9), 1980.

Fischer, J. E., Foster, G. S., Abel, R. M. et al., Hyperalimentation as primary therapy for inflammatory bowel disease, *Am. J. Surg.*, 125, 165, 1973.

Fischer, J. E., Rosen, H. M., Ebeid, A. M. et al., The effect of normalization of plasma amino acids on hepatic encephalopathy in man, *Surgery*, 80, 77, 1976.

Garcia, J. M., Mispireta, L. A., and Pinho, R. V., Percutaneous supraclavicular superior vena caval cannulation, *Surg. Gynecol. Obstet.*, 134, 839, 1972.

Glynn, M. F. X., Langer, B., and Jeejeebhoy, K. N., Therapy for thrombotic occlusion of long-term intravenous alimentation catheters, *J. Parenteral Enteral Nutr.*, 4(4), 387, 1980.

Land, R. E., Anatomic relationships of the right subclavian vein, *Arch. Surg.*, 102, 178, 1971.

Land, R. E., The relationship of the left subclavian vein to the clavicle, *J. Thorac. Cardiovasc. Surg.*, 63, 564, 1972.

MacFadyen, B. V., Jr. and Dudrick, S. J., Management of gastrointestinal fistulae with parenteral hyperalimentation, *Surgery*, 74, 100, 1973.

Moosman, D. A., The anatomy of infraclavicular subclavian vein catheterization and its complications, *Surg. Gynecol. Obstet.*, 136, 71, 1973.

Reilly, J., Ryan, J. A., Strole, W. et al., Hyperalimentation in inflammatory bowel disease, *Am. J. Surg.,* 131, 192, 1976.

Ryan, J., Abel, R. M., Abbott, W. M. et al., Catheter complications in total parenteral nutrition — a prospective study of 200 consecutive patients, *N. Engl. J. Med.,* 290(14), 757, 1974.

Scribner, B. H., Cole, J. J., Christopher, T. G. et al., Long-term total parenteral nutrition — The concept of an artificial gut, *JAMA,* 212(3), 457, 1970.

Wilson, J. N., Grow, J. B., Demong, C. V. et al., Central venous pressure in optimal blood volume maintenance, *Arch. Surg.,* 85, 563, 1962.

Wolfe, B. M., Keltner, R. M., and William, V. L., Intestinal fistula output in regular elemental and intravenous nutrition, *Am. J. Surg.,* 124, 803, 1972.

Chapter 6

POSSIBLE COMPLICATIONS ASSOCIATED WITH PARENTERAL NUTRITION ADMINISTRATION

Gail Kennedy

TABLE OF CONTENTS*

* Reading List follows Chapter 10.

I. INTRODUCTION

The complications of parenteral nutrition (PN) therapy may be divided into three categories: (1) septic, (2) metabolic, and (3) technical. There are numerous situations that may be classified into one or more of the above, but with proper patient monitoring, good observation, and good aseptic technique most complications are preventable. The following are the most common ones seen.

II. SEPTIC COMPLICATIONS

Sepsis is a serious complication of PN and patients receiving this therapy are susceptible to infection for various reasons. If a fever develops during the administration of PN, a close examination for focal sources must be carried out. In particular, injection abcesses, phlebitis, chest infection, urinary tract infection, and abdominal sepsis should be considered. The fever may also be associated with allergic reactions to the nutrients, an infected central venous line, contaminated fluid, or tubing.

If sepsis is suspected the patient should be carefully evaluated, and the following protocol pursued:

- Retrograde blood cultures (from the central venous catheter)
- Peripheral blood cultures
- Culture of all solutions being infused into the patient
- Culture of solution below the filter (if one is used)
- Cultures of wounds, drain sites, catheter insertion sites, urine, sputum, and throat
- Chest X-ray

The parenteral nutrition should be discontinued and the central venous line should be kept open by infusing an electrolyte solution appropriate for the needs of the patient. While waiting for cultures to incubate the patient is closely observed and vital signs monitored frequently. If the temperature returns to normal and remains so for 12 hr, the PN may be resumed through the same catheter. If the temperature does not return to normal and the source of the fever is not found in 12 hr, the catheter should be removed and cultures made from it. If PN is still required, a new catheter should be inserted.

The catheter should not be removed without first evaluating its role in regard to the febrile episode. Antibiotics may be subsequently necessary if the patient's temperature remains elevated, septicemia continues despite catheter removal, or if the cause of fever is a source of sepsis outside the catheter.

III. METABOLIC COMPLICATIONS

The details of metabolic complications are given in Chapter 3 under each nutrient. In addition the following problems may be observed.

A. Hyperglycemia

This will occur when the rate of infusion exceeds the rate at which the body can metabolize glucose. It may be caused by infusing the amino acid-dextrose mixture too rapidly, or by metabolic complications. The ability to utilize glucose may be decreased by trauma and sepsis. It should be noted that the sudden appearance of hyperglycemia in a patient who was previously euglycemic usually heralds an infection. This ability may also be compromised by the following conditions: diabetes, pancreatitis, liver disease, some antibiotics (cephalosporins), and steroids. The first sign of hyperglycemia is usually glucosuria (2+ or more Clinitest®). If this is allowed to continue and is not treated, it will lead to the development of osmotic diuresis, followed by hyperosmolar nonketotic acidosis and possibly death. The patient should be watched carefully for glucose intolerance and if glucosuria, dry skin, oliguria, confusion, or lethargy are noted, the nurse should be instructed to call the physician. If hyperglycemia persists due to continuing glucose intolerance, the blood glucose levels should be measured regularly and exogenous insulin may be infused or the flow rate decreased. Where insulin needs to be given, it is most economically given by a constant infusion for several reasons. If the patient is acutely sick and has low blood pressure or poor peripheral circulation then intravenous infusion (in contrast to subcutaneous administration) is the only way to ensure that injected insulin is available to the body. The intravenous route allows careful control of blood sugar in the unstable patient. Finally, the needs for insulin with a constant infusion are low compared with the amounts required using intermittent dosage.

B. Hypoglycemia

This is less common, but may occur if hypertonic glucose infusions flowing at a rapid rate are abruptly terminated or decreased. Mechanical causes, such as clogged filters, kinked tubing, or piggybacking additional medications may lead to this phenomenon. Symptoms include weakness, trembling, diaphoresis, headache, chills, rapid pulse, and decreased consciousness.

Attention to ensuring a constant flow of hypertonic glucose infusions will prevent hypoglycemia from occurring. If the volume infused falls behind schedule, do not increase the rate to "catch up". The rate should be recalculated in order to infuse at a uniform rate the prescribed amount of solution over the remainder of the 24 hr, provided this rate does not exceed 10% of the original drip rate.

C. Electrolyte Imbalances

The most common imbalances are related to the following electrolytes: potassium, phosphorus, and magnesium. During active protein synthesis and anabolism the level of the above ions in the plasma may fall. Careful monitoring of laboratory values and the general condition of the patient, i.e., urine output, heart rate, etc., will help to detect deficiencies and excesses. These should be corrected immediately, usually through a change in the PN prescription (see Chapter 3 for details).

D. Trace Element Deficiencies

Trace element deficiencies can be avoided by daily supplementation of these elements. Some hospitals are equipped to prepare their own trace mineral formulas (see Chapters 9 and 13), while others cannot and must treat the deficiencies symptomatically (see Chapter 3 for details).

E. Vitamin Deficiencies

Vitamins should be added to the PN daily, since most vitamin stores are depleted in malnourished patients. The input of fat-soluble vitamins should be carefully controlled; excessive intake of vitamins A and D can cause toxic effects, especially hypercalcemia (see Chapter 3).

F. Hyperlipoproteinemia

An elevation of blood lipids may occur as a result of overproduction of lipids from endogenous sources, overinfusion of exogenous lipids, or reduced utilization. Combinations of these factors may also be observed. The overproduction of lipids causing hyperlipidemia may result from infusing carbohydrates in excess of needs. Correspondingly, excessively high infusion rates of lipid emulsions may also cause hyperlipidemia. Under these circumstances reducing total caloric intake, both the carbohydrate and lipid moieties, can restore blood lipid levels to normal. Even with an appropriate caloric intake however, a deficiency of lipoprotein lipase, as with the syndrome of Type I hyperlipoproteinemia, may impair lipid clearance. This may occur also with severe protein deficiency and diabetes.

G. Hypercholesterolemia

This may be seen when lipid emulsions are infused at high rates, i.e., in excess of 2 g/kg/day. The rise is temporary and the level of plasma cholesterol returns to normal within a few days of discontinuing the infusion.

H. Abnormalities of Liver Function

The most common abnormality is a slight rise in the level of alkaline phosphatase, which returns to normal and has no clinical significance.

Fatty liver is a common problem too, due to the infusion of carbohydrate in excess of caloric needs (see Chapter 3 and references by Messing et al. for details).

Essential fatty acid deficiency also results in a fatty liver. Clinically the patient presents with a large tender liver and an elevation of the serum glutamic-oxaloacetic transaminase (SGOT).

The other hepatic clinical syndrome seen with TPN is cholestatic jaundice with elevated serum bilirubin and alkaline phosphatase. This syndrome occurs commonly in children and bears a relation to the duration of TPN. However, it is not seen frequently in home TPN patients despite the fact that they have been receiving TPN for years. Hence the condition does not result simply from prolonged TPN, but is probably due to a combination of prolonged TPN and acute illness. This concept of a combination of factors causing cholestasis is supported by the observation that the condition is largely seen in sick hospital patients rather than in relatively well patients receiving home TPN. One such factor may be sepsis since jaundice is most often seen clinically in adults who become septic or those with foci of infection.

I. Pancreatitis

When pancreatitis is seen in association with TPN it may potentially be due to either hypercalcemia or hyperlipidemia. However, in practice we have identified the former as being the main cause of pancreatitis (see reference by Izsak et al.).

J. Essential Fatty Acid Deficiency (EFAD)

EFAD is exhibited by dry scaly skin and hair loss. This can be prevented or corrected with administration of a fat emulsion intravenously. By using the lipid as a source of calories this deficiency will never be seen. For details see Chapter 3.

IV. TECHNICAL COMPLICATIONS

A. Clotted Catheter

If the catheter becomes clotted, an attempt should be made to unplug it. A 1-m*ℓ* syringe containing sterile normal saline is attached directly to the hub of the catheter and an attempt is made to loosen the clot by irrigation. If this fails, the clotted catheter should be unplugged by instilling urokinase (see reference by Glynn et al.). This enzyme will dissolve the fibrin in the catheter and allow it to become patent once again. The technique consists of dissolving 7500 IU of urokinase in 3 m*ℓ* of normal saline and injecting as much as will go into the catheter, up to a maximum of 2.5 m*ℓ*. Then the catheter is capped for 3 hr and flushed with a mixture of 10 m*ℓ* saline containing 1000 units of heparin.

B. Air Embolism

This is a potentially fatal complication. It can occur during the catheterization procedure, tubing change, or through disconnection of the tubing. Signs of air embolism are cyanosis, a rapid weak pulse, hypotension, and loss of consciousness. If air embolism is suspected, the patient is immediately placed in the Trendelenburg position on his or her left side so the air is confined to the right atrium, then further measures are taken to support the cardiovascular system.

Other complications that may be seen occasionally are catheter embolism, venous thrombosis, and suppurative thrombophlebitis. These can all be eliminated through good preventive nursing care and careful patient monitoring.

Chapter 7

MONITORING THE PARENTERAL NUTRITION PATIENT

Gail Kennedy

TABLE OF CONTENTS*

* Reading List Follows Chapter 10.

I. INTRODUCTION

Standardized procedures for monitoring should be developed and implemented in each hospital. These may include:

- Urine test for glucose and acetone every 6 hr (use Ketodiastix® for obtaining specific results as a false positive reading can occur with Clinitest®)
- Daily weight
- Vital signs four times daily
- Accurate intake and output records
- Laboratory tests

II. LABORATORY TESTS

These will vary with the patient's needs but some monitoring is vital. These tests can be divided into those which are absolutely necessary, those which monitor the nutritional status of the patient, and special optional tests.

A. Mandatory Tests

These include hemoglobin, white cell count, electrolytes, blood sugar, and blood urea nitrogen. Serum creatinine should also be monitored frequently. These observations should be made at least three times a week and selected ones more frequently as required for the control of unstable patients.

B. Tests to Monitor the Nutritional Status

These include serum albumin, serum iron and iron binding capacity (transferrin), serum calcium, magnesium, phosphorus, triglycerides, and cholesterol. In addition, estimation of transaminase levels, alkaline phosphatase, and bilirubin in the serum are important for recognizing the development of liver abnormalities. The tests given above should be done once a week.

C. Special Tests

These are optional and need be done as required. They include plasma zinc, copper, and fatty acid levels.

In addition to these purely laboratory parameters the patient must be monitored clinically as follows.

III. CLINICAL MONITORING

Sepsis: All principles of asepsis must be adhered to when dealing with any part of the PN protocol, i.e., changing the dressing, mixing the solutions, changing the container or tubing. Signs of infection at catheter insertion site, wounds, drains, etc. should be carefully looked for by the nurse.

Mouth care: This is very important in a physical and psychological sense; oral examinations are necessary to detect any complications, i.e., parotitis, glossitis, etc. Patients who are NPO may develop a dry tongue, throat, and mouth which may then become inflamed and uncomfortable. Oral care should include frequent use of mouth washes of various flavors, brushing of teeth, and use of lip balm. Patients may be allowed to chew gum or suck on hard candy.

Physical therapy: Patients should be encouraged not to remain in bed but be up for short, frequent walks, unless his or her condition contraindicates this. Physical therapy should be instituted in an active or passive form to promote protein synthesis and aid in the building of muscle mass.

Bowel function: Patients receiving PN may pass stools very infrequently. This change in bowel habit may bother some patients and they require reassurance that things will return to normal when eating resumes.

Appetite: If the patient is receiving enough calories he or she should not have any "hunger pains". There will be a decrease in appetite, although patients do think about food frequently. When the patient is allowed to resume eating the diet should be increased gradually. That is, by starting with clear fluids to determine tolerance, then on to mixed fluids given in five small feedings per day, and finally to a normal diet. It has been claimed that even when the GI tract has recovered, the patient receiving PN should be given an elemental diet during a transitional period prior to institution of normal food, so as to allow the bowel to adapt. This is quite unnecessary, and Greenberg et al. have shown that the bowel responds well to a meal fed abruptly after three weeks of bowel rest.

Psychological needs: Patients receiving PN are usually very sick. Prolonged illness and PN may be very stressful to both the patient and his or her family. They will require constant emotional support, reassurance, and answers to questions. Diversional activities may be necessary and in some cases even a consultation with a psychiatrist.

Chapter 8

INSERTION OF CENTRAL VENOUS CATHETER — NURSING ASPECTS

Gail Kennedy

TABLE OF CONTENTS*

* Reading List follows Chapter 10.

I. RATIONALE AND BACKGROUND

The main difficulty with infusing nutrients parenterally is the high osmolality of the nutrient solution, predisposing to venous thrombosis. This complication can be avoided by infusing these solutions into the superior vena cava where a high rate of blood flow rapidly dilutes the infused hypertonic solutions. This route is especially useful in patients requiring a prolonged infusion or those with poor peripheral veins. Furthermore, this approach allows the infusion of fluids irrespective of osmolality, is more comfortable for the patient, and avoids repeated venipunctures. The superior vena cava can be catheterized either by puncturing a central vein, such as the subclavian, and advancing the tip of the catheter into the cava, or by inserting a long catheter into the antecubital vein and advancing the tip to the appropriate location.

The most commonly used technique for cannulating the superior vena cava is subclavian vein puncture. The preferred side and access route is the left subclavian vein because its junction with the innominate vein is a gentle arch, allowing the catheter to slide in easily. However, the location of the thoracic duct makes the left side less advantageous and the fact that the apex of the left lung is higher than that of the right increases the chance of complicating pneumothorax. Peripheral long, plastic catheters inserted into the antecubital vein and advanced into the superior vena cava have produced a high incidence of thrombophlebitis, thus, their use is at present of limited value. Perhaps with the development of long silicone catheters, inflammation and thrombosis may decrease. Regardless, these types of catheters limit movement of the arm and present a problem in applying an air-occlusive dressing.

The jugular vein is another site for central venous catheterization. The external jugular is difficult to cannulate and is easily thrombosed. Internal jugular catheterization has a lower incidence of thrombosis, however both approaches lead to difficulties in maintaining an occlusive, sterile dressing at the site of the skin puncture. Most punctures for cannulating the internal jugular veins occur near or in the patient's hairline, which necessitates shaving this area to avoid introduction of any hair or bacteria into the insertion site. The patient also experiences limited head and neck movement, and the nurse must carefully monitor the flow rate because this is easily altered by neck movement.

Irrespective of the entry site, catheter placement should be performed or supervised only by experienced personnel because the complication rate is inversely proportional to the experience of the person inserting the catheter. Details of catheter placement are given in Chapter 5.

II. PATIENT PREPARATION

Prior to insertion of the catheter, the procedure should be thoroughly explained to the patient by the doctor or the nurse. If the patient does not understand English, an interpreter should be called. This will help to alleviate their fears and anxiety and perhaps lessen the discomfort experienced during catheterization.

The patient is placed supine in a Trendelenburg position (Figure 1) to provide positive pressure within the central vein and to promote filling and dilation of the subclavian vein. Each patient should be individually assessed as to their tolerance of this position, i.e., it is impossible if the patient is in cardiac or respiratory failure.

A small, rolled towel should be placed between the scapulae to allow the shoulders to fall back onto the bed. This increases the separation between the clavicle and first rib as well as separating the subclavian vein from the apex of the lung, thus allowing an easier entry. The patient's head should be turned away from the side of catheterization and rest flat on the bed (Figure 2)

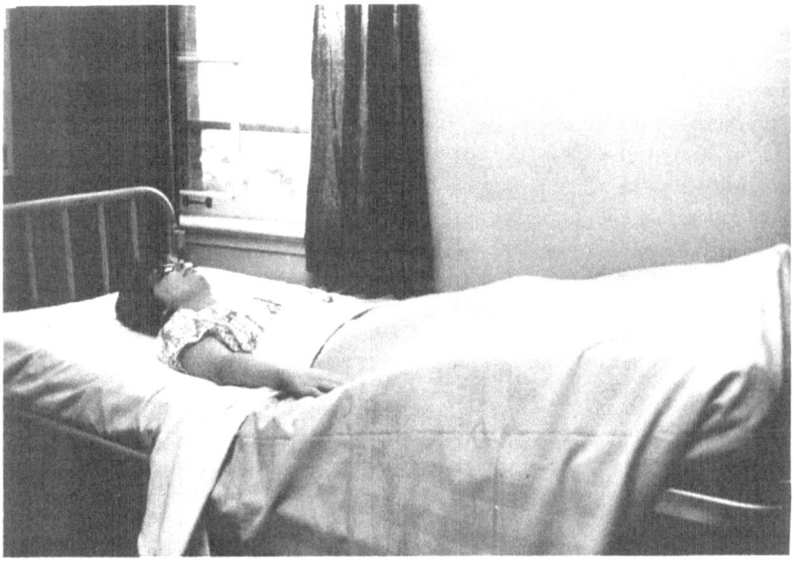

FIGURE 1. Supine and Trendelenburg positioning of patient for subclavian insertion of central vein catheter.

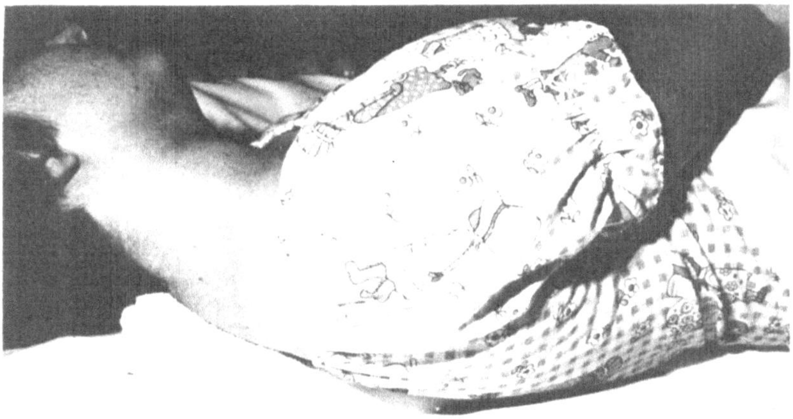

FIGURE 2. Further positioning with head turned away from side chosen for subclavian insertion of central vein catheter. Small rolled towel placed longitudinally between scapulae allows shoulders to fall back giving subclavian area exposure at a better angle.

The patient should be instructed in the Valsalva maneuver and should be allowed to practice it several times prior to insertion. This maneuver involves taking a deep breath, holding it, and bearing down (similar to the action with a bowel movement) which increases a positive pressure and reduces the chance of air entry and air embolism. Since this maneuver may be required three to four times during insertion it should be taught to the patient prior to inserting the catheter.

If the patient has any hair covering the insertion area, it should be shaved, then the skin defatted and prepared with a povidone-iodine solution and sterile drapes applied.

Asepsis must be maintained throughout without making the patient apprehensive. The draping and the noxious smell of the cleaning agents may lead to apprehension, thus reassurance and measures to ensure the patient's comfort are necessary.

Lidocaine, a local anesthetic, is injected into the subcutaneous tissue and causes a pressure sensation and stinging prior to the onset of anesthesia. This problem should be explained to the patient.

After thorough explanation and discussion with the patient, the need for premedication can be determined. The patient who is psychologically prepared and if necessary, sedated, will be more cooperative throughout the entire procedure.

III. CATHETER INSERTION

The nurse should be present to assist the physician and take an active role in comforting the patient. To accomplish this one must be totally familiar with the procedure described in Chapter 5, as well as being adept and knowledgeable about all equipment. Such knowledge will aid asepsis and also reassure the patient because it leads to a cooperative and methodical approach to the procedure.

A "prep tray" or "cart" is most advantageous, and if kept in patient areas where catheters are inserted will facilitate the whole procedure. If kept well stocked, supplies are accessible without the nurse having to leave the room. See Table 1 for a complete list of the equipment required.

Just prior to the insertion, a bag of normal saline is connected to the intravenous tubing which is flushed out with the saline solution. Then the patient is positioned. Everyone who is present in the room, including the patient, should wear a mask. It is at this point that the nurse's role becomes very important. The nurse should be able to anticipate the doctor's needs and yet support and comfort the patient.

After the catheter has been inserted it should be positioned in such a way as to lie flush on the chest parallel to the sternum. It should not be angled over the arm or shoulder. After arranging the catheter on the chest it should be sutured in place. The doctor or nurse can then place the initial catheter dressing.

Prior to dressing the catheter insertion site, remove any excess blood that has oozed from the puncture. The whole area should be cleaned with the defatting agent, followed by the application of povidone-iodine solution. When this solution has air-dried, povidone-iodine ointment is liberally applied to the catheter exit site, covered with a sterile 4″ × 4″ gauze, then an occlusive pressure dressing and Elastoplast® or waterproof tape is applied. This initial dressing should only stay in place for a maximum of 24 hr and the site should then be redressed according to one of several protocols outlined in the next section of this chapter.

Immediately after inserting the catheter, an isotonic intravenous solution should be infused slowly (to keep the vein open) and a chest X-ray taken and read to ensure that the catheter is properly situated in the superior vena cava and that no pneumothorax was created.

The nurse should be well acquainted with the signs and symptoms of the complications of the catheter insertion and watch for their occurrence (Chapter 5). Observations by competent nurses are invaluable in the recognition of possible complications during and after catheter insertion, as well as throughout the duration of parenteral nutrition therapy.

IV. CENTRAL CATHETER CARE — DRESSINGS

Care of this "nutritional lifeline" is a very important function of the nurse's role. The principles of asepsis should be clearly understood and practiced to prevent con-

Table 1
PREP TRAY OR CART

Modified Intravenous Cut Down Tray

1 small K basin	1 stitch scissors
1 instrument basin	1 curved snaps
1 beaker	2 Strabismus hooks
2 medicine glasses	1 disposable syringe, 2 mℓ
1 Kocher forceps	1 disposable syringe, 20 mℓ
1 plain tissue forceps	2 scalpel handles, #3
1 toothed tissue forceps	2 scalpel blades, #15
1 small curved scissors	1 needle holder
1 small straight scissors	1 curved needle with black silk (swag on)
1 disposable needle, 22ga. 1½"	
1 disposable needle, 25 ga. 5/8"	
2 Towels	

Other Items

Masks
Defatting agent (benzalkonium chloride 1:750, alcohol, acetone)
Povidone-iodine solution 10%
Povidone-iodine ointment (antibacterial and antifungal)
Sterile towels
Sterile gloves (various sizes)
Local anesthetic (usually lidocaine 1% or 2% without epinephrine)
Extra syringes (2 and 5 mℓ)
Extra needles (21 and 25 gauge)
Suture material (4-0 monofilament nylon skin sutures)
Intracatheters (16 gauge, 8 in. long)
Gauze (4" × 4")
Cotton swabs
Sterile dressing trays
  2 Containers for cleaning solutions
  Forceps
  Sterile towel
Adhesive tape (1.25-cm wide)
Waterproof or Elastoplast® tape
Extension tubes
Straight intravenous tubing
Isotonic intravenous solution (500 mℓ 0.9% normal saline or 5% dextrose in water)
Unsterile towel
Underpad
Garbage bags
Intravenous pole

tamination or sepsis at the site of line insertion. The type of dressing utilized will depend on the preferences of the hospital, the doctor, and nursing service. In addition it may be dictated by the patient's condition, i.e., increased perspiration, exfoliative skin problems, etc. It is important to stress that each hospital must establish a standardized procedure to change the dressing. The author prefers an air-occlusive, waterproof dressing called Op-Site®.

The type of dressing chosen will dictate the frequency of change. For example, a dressing with Op-Site® (a waterproof, hypoallergenic, plastic material) usually needs to be changed only once a week, but may have to be changed earlier if it becomes loose or soiled. In contrast, a gauze dressing covered by a waterproof tape requires more frequent dressing changes due to loosening of the tape, skin irritation, and the need to view the catheter exit site normally covered by gauze in a standard dressing.

To reduce the possibility of air-borne contamination a mask should be worn by the nurse performing the procedure, the patient, as well as by anyone else in attendance.

The use of sterile gloves is a controversial issue. Hospital policy may dictate the use of gloves. In any case the nurse should be fully adept in no-touch technique with sterile instruments. If gloves are not worn, the hands should be scrubbed with povidone-iodine. Some institutions may also insist that the nurse wear a cap and gown when changing the dressing.

To aid in the dressing change a sterile dressing tray is required. There are several options available: one is to order a custom-made sterile disposable tray through a medical supply company. Everything that the nurse requires for the change should be packed in the tray. If this option proves to be too expensive or impractical a hospital-packed sterile stainless steel tray with solution cups or bowls, clamps, forceps, and sterile drape is easily adapted to meet the nurse's needs. In addition, sterile 4″ × 4″, 2″ × 2″, and ½″ × ½″ gauze sponges, cotton balls, cleansing solutions, and povidone-iodine ointment are placed in the tray using aseptic technique.

Whatever choices are made the following principles and technique should be observed.

V. BASIC DRESSING

1. Explain the procedure to the patient to minimize his or her fears and enhance cooperation.
2. Place the patient in a supine position with head turned away from the dressing site. This will help prevent the catheter from accidentally falling out when the dressing is removed and reduces the risk of air embolism.
3. Scrub hands with an antiseptic solution, i.e., povidine-iodine, and also clean the surface where the tray will be opened and prepared.
4. Prepare tray if necessary, i.e., add defatting agent, povidone-iodine solution, povidone-iodine ointment, gauze sponges, cotton balls.
5. Remove old dressing carefully so as not to dislodge the catheter or contaminate it.
6. Observe site and surrounding skin for signs of inflammation, swelling, or discharge. Check the condition of the sutures maintaining the placement of the catheter. Any adverse symptoms should be noted, recorded, and a doctor informed.
7. Swab the catheter insertion site and send for culture and sensitivity. This will help to determine if the dressing is satisfactory.
8. Rewash hands thoroughly.
9. Defat the skin, i.e., use benzalkonium chloride 1:750 or alcohol. This is an important step, for it removes excess skin oils and fat, old debris, and destroys the integrity of the bacterial cell walls, while popular, harsh defatting agents like acetone should be used with care because they remove natural skin oils, resulting in skin irritation, cracking and breakdown, and secondary infection. This agent may also cause plastic and some silicone catheters to weaken and disintegrate.
10. An area of 10-cm radius around the catheter should be cleaned using a circular motion, moving from a clean area to the dirty one, i.e., from the catheter exit site outwards. A new cotton ball should be used for each circular swabbing action. Remember to clean under the needle guard because bacteria and old blood are likely to collect there. The final swab should start at the insertion site, cleaning vertically over the needle guard to the needle hub.
11. Repeat this procedure with an antiseptic such as povidone-iodine, which should be left to act in the wet state for at least 2 min.
12. Leave a thin film of povidone-iodine solution on the skin. If there is excess solution, mop it up with a sterile 4″ × 4″ gauze to ensure that the skin is dry.

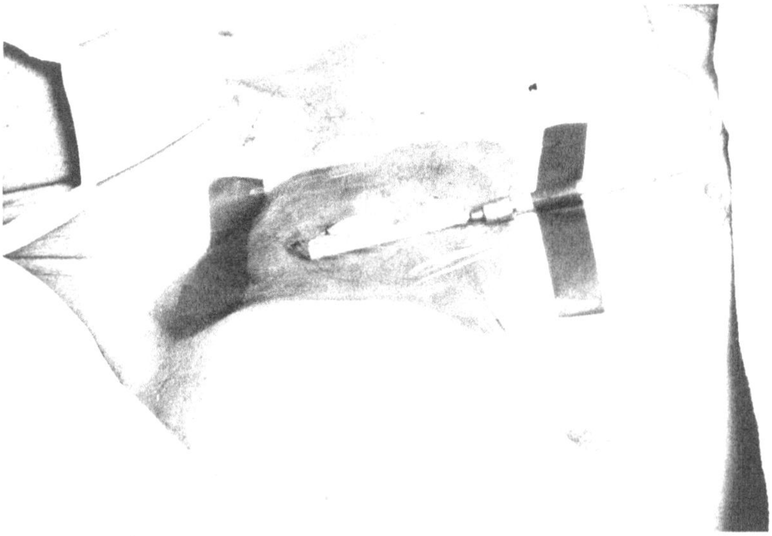

FIGURE 3. Op-Site® used as final dressing, in place over site of catheter exit and connection. Note patient is masked.

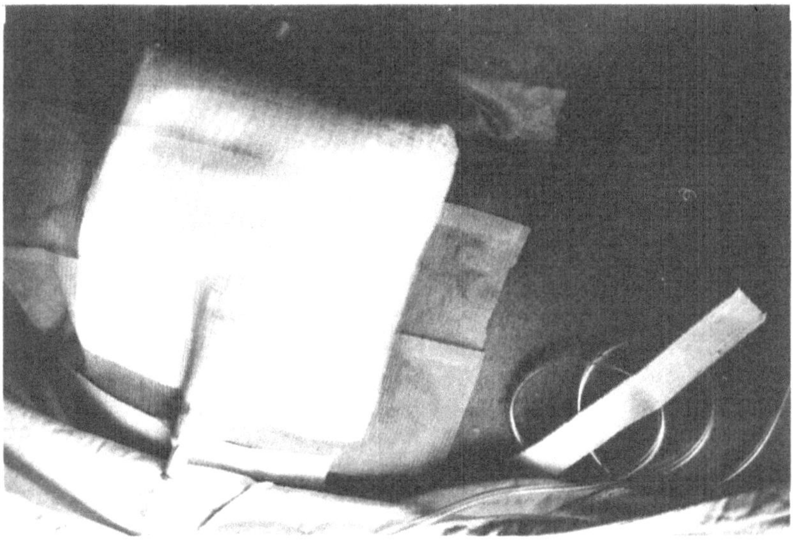

FIGURE 4. Gauze used as final dressing, in place over site of catheter exit and connection. Note adjacent coils of tubing taped to chest act as a buffer for undue tugs upon tubing.

13. Apply a small amount of povidone-iodine ointment only to the puncture site. This ointment is preferred because of its broad range against bacteria and fungi. Use of antibiotic ointments has been shown to increase the incidence of fungal colonization of the catheter.
14. A small ½" × ½" gauze, or slightly larger gauze, should be placed over the ointment. This keeps the ointment localized to the catheter exit site and prevents melted ointment from running down the catheter.
15. Application of the final dressing cover is done by two techniques.

- Op-Site® technique (Figure 3). With this technique it is important to locate the catheter insertion site 2.5 cm below the green edge of the Op-Site® so as to ensure that the total catheter will be covered, including the needle hub, up to the needle adapter. Apply the Op-Site® to itself around the adapter.
- Gauze technique (Figure 4). If a gauze dressing is preferred, one or two pieces of gauze should cover the catheter and then an occlusive tape, such as waterproof Band-aid® tape or adhesive-backed elastic tape should be used. If the tape does not adhere well, spray the area to be taped with tincture of benzoin.

16. Application of either of the above dressings should be done with the patient's arm abducted and head rotated away from the dressing. This allows the dressing to remain intact despite a full range of shoulder movement.
17. Secure the connection between the intravenous tubing and needle adapter with tape. This is the weakest connection in the whole tubing system and the tape prevents tubing disconnection and the possibility of air embolism.
18. Coil the extension tube or excess intravenous tubing laterally away from the dressing and tape it to the chest so that it lies beside the neck and just below the shoulder. This coil will absorb the shock of any tugging or pulling on the intravenous tubing thus protecting the catheter (Figure 4).
19. Mark on the dressing the date and initials of the nurse who changed the dressing.
20. Record the performance of this procedure, with pertinent remarks, on the appropriate hospital document.
21. Observe the dressing at each shift. Look for any signs of hypersensitivity to the dressing and for loose, contaminated, or wet dressings. If these occur the dressing should be changed promptly.
22. At the Toronto General Hospital, Op-Site® has been used exclusively for the past several years. On the basis of this experience the author highly recommends this type of covering for several obvious reasons. Since it is transparent it allows easy observation of the catheter and the skin puncture site at all times, thus any problem at the catheter exit site and along its extent is easily visible. This dressing only requires one change per week, saving much time for the nurse and aiding patient comfort.
23. If applied correctly it is bacteria-proof as well as waterproof, yet allows the skin to breathe. There is virtually no skin reaction to this material, since it is hypoallergenic. It is not bulky and looks very neat on the patient's chest.
24. With proper care and a conscientious approach many complications, especially infections, associated with central venous catheters can be eliminated. It is through insistence on good aseptic technique, updating of procedures, inservice education, and careful monitoring of the patient that a high quality of patient care is ensured.

Chapter 9

ORDERING, PREPARATION, AND INFUSION OF THE PARENTERAL NUTRITION SOLUTION

Gail Kennedy

TABLE OF CONTENTS*

* Reading List follows Chapter 10.

I. INTRODUCTION

The physician is responsible for prescribing the nutritional formula for the patient. It will vary with each patient, depending on individual requirements and results of monitoring by laboratory tests, i.e., electrolyte levels, serum protein levels, etc.

II. PATIENT PREPARATION

Prior to starting TPN, the following points should be considered:

1. The patient should be stabilized in regard to fluid and electrolyte status. Dehydration and electrolyte deficits should be corrected as the initiation of TPN may exacerbate any potassium and magnesium deficiency. These electrolytes are taken up by cells under the anabolic drive of the infused nutrients and associated rise in insulin levels. Furthermore, diuresis caused by inadvertent hyperglycemia may increase any existing dehydration. On the other hand, patients with edema and fluid overload should be given diuretic treatment before starting TPN. The TPN infusion results in an additional fluid load which may induce heart failure and pulmonary edema in patients who are already fluid overloaded.
2. It is also desirable to transfuse patients with blood and plasma in an effort to raise very low hemoglobin or albumin levels. Initially, just after starting TPN and prior to the nutritional support taking effect there is often a tendency for the hemoglobin and plasma protein levels to fall, which may be undesirable in severely depleted patients.
3. It should be recognized that TPN adds a fluid and electrolyte load which may result in an undesirable excess in relation to needs in patients who are already receiving other infusions such as antibiotics, pressor agents, and steriods in intensive care units. For example, the giving of the antibiotic carbenicillin may result in the infusion of as much as 30 g of sodium per day, combined with the antibiotic as a salt. All these sources of fluids and electrolytes should be calculated and the sum total added to that in the TPN to arrive at a figure which may be suitable for the patient. In such patients it may be necessary to use electrolyte-free amino acid mixtures to balance the intake from these other sources.
4. The presence of renal and hepatic failure should be considered in regard to amino acid tolerance and the need for additional frequency of dialysis.
5. Diabetes and glucose intolerance is another area that should be thought of, so as to foresee the need for insulin and ensure the patient is able to tolerate large carbohydrate loads.
6. The patient's history of septic foci, current sources of infection, and existing temperature should be noted so that the subsequent occurrence of fever can be considered in the light of the initial clinical condition.
7. Finally, any bleeding diathesis or impediments to catheter placement should be considered.

III. PROTOCOLS

When formulating protocols it is necessary to avoid the following mistakes:

1. Prescribe a balanced intake of amino acids, energy, electrolytes, vitamins, and trace elements. There is no place for bizarre prescriptions, such as the infusion of one bottle of Intralipid® alone, under most circumstances.

2. Carefully consider the electrolyte and acid-base status of the patient in relation to the prescribed formula. In particular, do not forget that the patient may be receiving other sources of electrolytes and H^+. Do not prescribe incompatible salts, such as sodium bicarbonate with calcium chloride. For further details about incompatibilities see Chapter 17 prior to formulating a prescription.
3. The concentration of macro- and micronutrients should be considered. For example, a prescription resulting in a final concentration of 20% amino acids and 70% dextrose is not physically possible. Also, the possible concentrations are restricted by commercially available products and limitations of flow through the catheter. Check the prescriptions for the final concentration of macro- and micronutrients.
4. Remember that each unit of TPN should be infused within a reasonable time, not exceeding 12 hr, to ensure a reasonable flow rate through the catheter. A very slow flow may result in clotting of the catheter. Also, some nutrients such as vitamins should not be allowed to remain in the bottles for long periods while infusing, as they are subject to decomposition.

In the following subsection a few representative standard protocols are presented. They can be a starting point for individually tailored mixtures or they may be used as such in adult patients who are malnourished and who do not have special requirements.

A. Representative Protocols for the Administration of TPN

The following protocols are designed to administer 1 g of amino acids with 40 nonprotein calories per kilogram of body weight per day, and are given for central infusion as (1) dextrose alone, (2) a dextrose-lipid mixture in which half the calories are as dextrose and the other half as lipid, and (3) for peripheral infusion as a mixture in which 27% of the calories are provided as dextrose and 73% as lipid. Quantities noted are based on the average 24-hr intake for an average 60 kg human.

1. Central System

Dextrose Formula
"100% glucose" (nonprotein calories)

0700 hr to 1900 hr

2% Amino acids and 25% dextrose	1500 ml
Electrolyte mix[b]	30 ml
M.V.I.-12[d,n]	5 ml 1 day/week[o]
Vitamin A 2500 IU/ml[f]	1 ml daily
Solu-Zyme® with Vitamin C[e]	10 ml daily 6 day/week
Synkavite[i]	10 mg once weekly

1900 hr to 0700 hr

2% Amino acid and 25% dextrose	1500 ml
Electrolyte mix[b]	30 ml
Potassium phosphate[b]	6 ml
Trace elements[m]	1 ml (each)

Total intake

Electrolytes (mmol)	
Na	137
K	78

Ca[c]	10.5
Mg	12
Cl	198
Acetate	114.6
PO_4 (P mg)	12 (390)
Trace elements (mg)	
Cu	0.2
I	0.12
Z	3.0
Mn	0.7
Cr	0.02
Se	0.12
Fe	1.0
Vitamins[d,e,f,n]	
Fluid volume	3000 mℓ (approx.)
Protein equivalence[h]	63 g
Nitrogen[g]	10.08 g
Carbohydrate	750 g
Kilocalories	
Protein[i]	258.3
Carbohydrate[k]	2550.0
	2808.3 kcal
Nitrogen:calorie ratio	1:278.6

Dextrose-Lipid Formula
"50% glucose and 50% lipid" nonprotein calories

0700 hr to 1900 hr

4.2% Amino acids and 25% dextrose	750 mℓ
Electrolyte mix[b]	30 mℓ
M.V.I.-12[d,n]	5 mℓ 1 day/week[o]
Vitamin A 2500 IU/mℓ[f]	1 mℓ daily
Solu-Zyme® with Vitamin C[e]	10 mℓ daily 6 day/week
Synkavite[f]	10 mg once weekly
Lipid 10%[j]	500 mℓ

1900 hr to 0700 hr

4.2% Amino acids and 25% dextrose	750 mℓ
Electrolyte mix[b]	30 mℓ
Potassium phosphate[b]	6 mℓ
Trace elements	1 mℓ (each)
Lipid 10%[j]	500 mℓ

Total intake

Electrolytes (mmol)	
Na	137
K	78
Ca	10.5
Mg	12
Cl	199
Acetate	117
PO_4 (P mg)	12 (390)
Trace elements (mg)	
Cu	0.2
I	0.12
Zn	3.0
Mn	0.7
Cr	0.02

Se	0.12
Fe	1.0
Vitamins[d,e,f,n]	
Fluid volume	2500 mℓ (approx.)
Protein equivalence[h]	66.2
Nitrogen[g]	10.6 g
Carbohydrate	375 g
Lipid	100 g
Kilocalories	
Protein[i]	271.4
Carbohydrate[k]	1275
Lipid[j]	1100
	2646.4
Nitrogen:calorie ratio	1:249.7
% Nonprotein calories as lipid	46.3

2. Peripheral Systems[a]

Peripheral Formula
"73% lipid and 27% glucose" nonprotein calories

0700 hr to 1900 hr

6% Amino acids with 12% dextrose	500 mℓ
Electrolyte mix	30 mℓ
M.V.I.-12[d,n]	5 mℓ 1 day/week[o]
Vitamin A 2500 IU/mℓ[f]	1 mℓ daily
Solu-Zyme® with Vitamin C[e]	10 mℓ daily 6 day/week
Synkavite[f]	10 mg once weekly
Lipid 10%[j]	500 mℓ

1900 hr to 0700 hr

6% amino acids and 12% dextrose	500 mℓ
Electrolyte mix[b]	30 mℓ
Potassium phosphate[c]	6 mℓ
Trace elements[m]	1 mℓ (each)
Lipid 10%[j]	500 mℓ

Total intake

Electrolytes (mmol)	
Na	137
K	78
Ca	10.5
Mg	12.0
Cl	198
Acetate	114.6
PO_4 (P mg)	12 (390)
Trace elements (mg)	
Cu	0.2
I	0.12
Zn	3.0
Mn	0.7
Cr	0.02
Se	0.12
Fe	1.0
Vitamins[d,e,f,n]	
Fluid volume	2000 mℓ (approx.)
Protein equivalence[h]	63 g

Nitrogen[g]	10.08 g
Carbohydrate	120 g
Lipid	100 g
Kilocalories	
Protein[i]	258.3
Carbohydrate[k]	408.0
Lipid[j]	1100.0
	1766.3
Nitrogen:calorie ratio	1:175
% Nonprotein calories as lipid	73.0

[a] Peripheral system can also be used centrally.

[b] Total electrolytes have been calculated on the assumption that the "electrolyte-free" amino acid solution is used with 60 m*l* of electrolyte mix and 6 m*l* of potassium phosphate, as manufactured by the Toronto General Hospital Pharmacy. For patients with excessive fluid losses or fluid retention, appropriate changes should be made. Electrolyte mix (mmol/20 m*l*): Na^+ 45, K^+ 20, Ca^{++} 2.5, Mg^{++} 4, Cl^- 58, $Acetate^-$ 20.8, Potassium phosphate (mmol/m*l*): PO_4^{--} 2.1, K^+ 3.0.

[c] Calcium added as calcium gluconate (2.25 mmol/*l*).

[d] M.V.I.-12® (USV Laboratories, Tuckahoe, N.Y.), 5 m*l* contains vitamin C 100 mg, vitamin A 3300 IU, vitamin D 200 IU, vitamin E 10 IU, thiamine HCl 3.0 mg, riboflavin 3.6 mg, niacinamide 40 mg, pyridoxine HCl 4 mg, D-pantothenic acid 15 mg, biotin 60 μg, folic acid 400 μg, vitamin B_{12} 5 μg.

[e] Solu-Zyme® with Vitamin C (Upjohn Co., Kalamazoo, Mich.) 1 vial (about 10 m*l*) contains vitamin B_{12} 25 μg, folic acid 5 mg, thiamine HCl 10 mg, riboflavin 10 mg, pyridoxine HCl 5 mg, D-pantothenic acid 45 mg, niacinamide 250 mg, vitamin C 500 mg.

[f] Vitamin A 2500 IU, can be manufactured by diluting Aquasol A Parenteral®, 50,000 IU/m*l* (USV Laboratories, Tuckahoe, N.Y.).

[g] Amino acid used for the calculation is Travasol® 10%, containing 1.68 g of nitrogen/*l*.

[h] Each g of nitrogen is equivalent to 6.25 g of protein.

[i] Each g of protein provides 4.1 kcal.

[j] Nutralipid® 10% (Pharmacia, Montreal, Canada) or Liposyn® 10% (Abbott Laboratories, Chicago): each m*l* of emulsion provides 1.1 kcal.

[k] Each g of dextrose monohydrate provides 3.4 kcal.

[l] Synkavite (Roche Laboratories, Nutley, N.J.): menadiol sodium diphosphate 10 mg/m*l*.

[m] Single-entity trace elements are manufactured by the Toronto General Hospital. There are six formulations: (1) Cu^{++} 0.2 mg/m*l*; (2) I^- 0.12 mg/*l*; (3) Zn^{++} 3 mg/m*l*; (4) Mn^{++} 0.7 mg/m*l*; (5) Cr^{+++} 0.02 mg/m*l*, Se^{++++} 0.12 mg/m*l*; and (6) iron-dextran 1.0 mg/m*l*.

[n] To provide minimal amounts of vitamin D. See Chapter 3 for the possible complications of hypercalcemia and metabolic bone disease observed with use of vitamin D, even in the small doses present in M.V.I. formulations.

[o] The author recommends the use of vitamin A alone for short-term TPN. However, if it is desired to give vitamin D then M.V.I.-12 may be substituted on 1 day of the week instead of vitamin A.

Hospitals that have a large number (30 to 40) of patients receiving parenteral nutrition require that these orders reach the pharmacy department by a certain hour in the day. This aids the pharmacy in planning their workload and meeting deadlines. The nurse should ensure that a carbon copy of the Order Sheet (Tables 1A and 1B) reaches the pharmacy department and that the original remains in the patient's chart. The nurse may transcribe these orders onto a work sheet. In any case, the nurse should refer to the original order to check the final preparation sent from the pharmacy.

IV. PREPARATION OF THE AMINO ACID-DEXTROSE SOLUTION

The actual preparation of solutions should occur in a low traffic area that is very clean, with good ventilation, and in a laminar flow hood. The mixing should be done in the Pharmacy department by personnel using strict aseptic technique (see Chapter 14).

There are various systems of preparation and hospitals must assess their own situation and choose the method that best meets their facilities and needs. Hospitals with limited facilities and a small demand may use the kit method (see Chapter 14); other

Table 1A
TORONTO GENERAL HOSPITAL
TOTAL PARENTERAL NUTRITION ORDER SHEET

Diagnosis: ______
Reason for TPN: ______
Initial Weight: ______ Height: ______
Subclavian: ______ Peripheral: ______
Date Ordered: ______ Date and Time TPN to be Started: ______
Date Order Changed: ______
REMARKS: ______

PATIENT'S ADDRESSOGRAPH

		PRESCRIBED TIMES (hrs) 0700 - 1900	1900 - 0700
AMINO ACID AND DEXTROSE	Amino Acids 2.1% and Dextrose 25%	______ ml	______ ml
	Amino Acids 4.2% and Dextrose 25%	______ ml	______ ml
	Amino Acids 6% and Dextrose 12%	______ ml	______ ml
	Amino Acids __% and Dextrose __%	______ ml	______ ml
	Others: ______	______ ml	______ ml
	______	______ ml	______ ml
LIPID:	FAT EMULSION 10% WITH 500 UNITS OF HEPARIN PER 500 ml OF FAT EMULSION	______ ml	______ ml
ADDITIVES: (Note: See precautions on reverse side)			
a) ELECTROLYTES	Electrolyte Mix	______ ml	______ ml
	Calcium gluconate (1 mEq = 0.5 mM)	______ mM	______ mM
	Potassium Chloride (1 mEq = 1 mM)	______ mM	______ mM
	Magnesium Sulphate (1 mEq = 0.5 mM)	______ mM	______ mM
	Sodium Chloride (1 mEq = 1 mM)	______ mM	______ mM
	Sodium Lactate (1mEq = 1 mM)	______ mM	______ mM
	Potassium Phosphate	______ mM	______ mM
	Others: ______	______	______
b) TRACE ELEMENTS:	1. Cu	______ mg	
	2. I	______ mg	
	3. Zn	______ mg	
	4. Mn	______ mg	
	5. Cr	______ mg	
	6. Se	______ mg	
	7. Fe (Iron Dextran)	______ mg	
c) VITAMINS:	M.V.I. ______ day(s) per week	______ ml	
	Solu-Zyme with Ascorbic Acid ______ days per week	______ ml	
	Synkavite per week	______ mg	
	Vitamin A Inj. (2500 iu/ml) daily	______ IU	
d) ADDITIONAL MEDICATIONS:	1) ______	DOSE ______	
	2) ______	DOSE ______	

24 HOUR INTAKE

Electrolytes (mM)	Trace Elements (mg)	K Calories	Fluid Volume	Total Volume
Na ______	Cu ______	LIPID ______ } Total K Cal ______	Amino Acids and Dextrose ______ ml }	______ ml
K ______	I ______	CHO ______	Lipid ______ ml	
Ca ______	Zn ______	PROTEIN ______		
Mg ______	Mn ______	PROTEIN ______ gm		
Cl ______	Cr ______	CHO ______ gm		
HCO_3 ______	Se ______	FAT ______ gm		
Acetate ______	Fe ______	NITROGEN (N_2) ______ gm		
Lactate ______				
PO_4 ______				

______ NURSE ______ PHARMACIST ______ DOCTOR

T.G.H. FORM 299 (Rev. 8.81) PINK — Patients Chart Copy/YELLOW — Pharmacy Copy

hospitals may use standardized solutions which are modified to meet individual patient requirements, while yet others may prepare each patient's order every day. For specific patient problems, such as those with renal, cardiac, or hepatic disease, special formulas may be compounded. In larger hospitals where the demand for both standard and individually tailored solutions is high, the solutions are made in bulk and aseptically transferred to sterile containers (see Chapter 14).

In any case it is conventionally enunciated that the entire requirements for a patient should be mixed only in the pharmacy, with all additives mixed at the time of compounding. The ready-made individual containers are then sent to the ward where the role of the nurse is to check that the orders on the chart correspond to the label on

Table 1B
PROTOCOL

ROUTINE T.P.N. PROTOCOL

1. Prior to T.P.N. administration, review and follow T.P.N. protocol as outlined in the Manual. Notify T.P.N. Teacher (locating 3155).
2. Routine lab work before T.P.N., during T.P.N., after T.P.N. Tests printed in bold type done as ordered. Tests printed in regular type done when checked by M.D. Stamp requisition with T.P.N. stamp.
3. PRECAUTION: Check for medication compatibilities with pharmacist when additional medication and electrolytes are to be added to solutions. Certain electrolytes and drugs may interact and precipitate out (local 3611).
4. Diabetic Protocol q.i.d.
5. Weights daily.

BEFORE T.P.N. AND AFTER T.P.N.	DURING T.P.N.
Blood: BUN BS Creat Elec Hgb Hct WBC Diff Ca P Mg Protein A/G Ratio SGOT Cholesterol Triglycerides Serum Fe B 12 Folate Bilirubin T & D Alk Phos Protime Cultures and Candida: Throat Urine Sputum Wound Swabs Nasal Swab Repeat above at End-Alimentation Chest X-Ray	Weekly: (Monday) **HCT WBC Protein A/G Ratio** Creat Ca P Mg Hgb Bilirubin T & D SGOT Alk Phos Se Fe and IBC, **Triglycerides & Chol.** 2 × weekly: (Monday and Thursday including holidays) **BUN BS Elec** **Weekly Cultures** **Catheter insertion site** **All wound drainage**

T.P.N. INFUSIONS	UNIT VOLUME (ML)	*PROTEIN EQUIVALENCE (GM)	CHO (GM)	FAT (GM)	K CALORIES ØØ PROTEIN	K CALORIES ** NON-PROTEIN	K CALORIES TOTAL	Ø NITROGEN (GM)	ELECTROLYTES (mM) PER UNIT VOLUME Na^+	K^+	Ca^{++}	Mg^{++}	Cl^-	Acetate	Lactate	PO_4
a) STANDARD SOLUTIONS OF AMINO ACIDS AND DEXTROSE																
Amino Acids 4.2% and Dextrose 25%	1000	44.10	250		180.8	850	1030.8	7.05	1.3	12.0	2.25		16.8	36.5		8.4#
Amino Acids 4.2% and Dextrose 25%	500	22.05	125		90.4	425	515.4	3.52	0.66	6.0	1.13		8.4	18.25		4.2#
Amino Acids 6% and Dextrose 12%	1000	63.0	120		258.3	408	666.3	10.08	1.8	12.0	2.25		24.0	52.2		8.4#
Amino Acids 6% and Dextrose 12%	500	31.5	60		129.2	204	333.2	5.04	0.9	6.0	1.13		12.0	26.1		4.2#
Travasol 10% Electrolyte Free	1000	105.0			430.5		430.5	16.80	3.0				40.0	87.0		
b) ELECTROLYTES																
Electrolyte Mix — TGH	20								45.0	20.0	2.5	4.0	58.0	20.8		
Calcium Gluconate 10%	10										2.25					
Magnesium Sulphate 50%	2											4.0				
Potassium Chloride Inj. USP (40 mEq/20 ml)	20									40.0			40.0			
Potassium Chloride Inj. USP (20 mEq/10 ml)	10									20.0			20.0			
Sodium Chloride 23.4% T.G.H.	20								80.0				80.0			
Sodium Lactate 44.8% T.G.H.	20								80.0						80.0	
Potassium Phosphate K 3 mM/ml P 2.1 mM/ml	30									90.0						63.0
c) FAT EMULSION																
Nutralipid 10% (Intralipid)	500			50		550		0.105	0.4	0.13	0.01	0.002				6.8-7.4
Liposyn 10%	500			50		550		Not Available	0.4	0.13	0.01	0.002				6.8-7.0

d) VITAMINS: (Conc/ml)	
M.V.I. - 1000	Vit. C 100 mg, Vit. A 1000 I.U., Vit. D 100 I.U., Vit. E 1 I.U., Thiamine HCl 4.5 mg, Riboflavin 1 mg, Niacinamide 10 mg, Pyridoxine HCl 1.2 mg, d-Panthothenic Acid 2.6 mg.
Vit. A. Inj.	Vit. A 2500 I.U.
Synkavite	Menadiol Sodium Diphosphate 10 mg.
Solu-zyme (Conc/vial)	Vit. B_{12} 25 mcg, Folic Acid 5 mg, Thiamine HCl 10 mg, Riboflavin 10 mg, Pyridoxine HCl 5 mg, d-Pantothenic Acid 45 mg, Niacinamide 250 mg, Vit. C 500 mg.
Lipid (Conc/L)	Vit. E I.U.: Nutralipid (Intralipid) 10% approximately 30-35, Liposyn 10% approximately 30.
e) Trace Elements:	Dose in uncomplicated Patients (mg/day): Cu 0.3, I 0.12, Zn 3.0, Mn 0.7, Cr 0.02, Se 0.12, Fe 1.0

* Calculated using empirical formula, each gram of N_2 is equivalent to 6.25 grams of protein.
** Each gram of Dextrose Monohydrate provides 3.4 K Cal.
Ø Calculated using 10 grams of amino acids of Travasol 10% containing 1.68 grams of Nitrogen.
ØØ Each gram of protein equals 4.1 K Cal.
\# Assumption that phosphate is in divalent form, thus each mM of $PO_4^{=}$ contains 31 mg of P.

the container, prior to administration. It should also be noted that once the amino acid solution is mixed with dextrose and various electrolytes and minerals it is recommended that they be kept refrigerated to ensure optimum safety. Furthermore, when vitamins and other medications are added, they should be used within 24 hr.

The above protocol is followed to ensure that solutions are not contaminated on the ward by additives injected into the solution in a haphazard way. This method also increases the load on the pharmacy in hospitals with a large number of parenteral nutrition patients.

In contrast to the above, staffers at the Toronto General Hospital have practiced a technique which divides the responsibility between pharmacy and nursing. This system has been in operation for over ten years and is to be recommended for its simplicity, safety, and cost-effectiveness. At the Toronto General Hospital the pharmacy prepares 100 to 150 ℓ of standard solutions (Chapter 14) in a large stainless-steel tank and sterilizes the fluid by passing it through two filters. The fluid is pumped into a plastic bag (TA-10, Transfer Pack, Fenwal®) to a predetermined weight to provide 1 ℓ or 500 mℓ. Plastic bags are preferred for a number of reasons. They require less storage space as they can be stacked, they collapse as the fluid empties thus decreasing the possibility of air embolism and if they are dropped or fall from an intravenous pole there is no danger to the patient or the nurse from broken glass.

The required number of bags of standard solution is regularly stocked in the designated nursing units, with an expiration date of 60 days (or 30 days in the case of individually prepared solutions which are not sterilized by membrane perfusion). These solutions are kept refrigerated until 1 hr prior to use. The nurse then removes the solution to allow it to reach room temperature prior to administration, as infusion of cold solutions could cause vasoconstriction and change body temperature.

When the solution is out of the refrigerator the nurse should assemble all supplies, i.e., additional additives, syringes, needles, and alcohol swabs. This should be done in a clean, quiet area, such as the medication room or clean utility room. The surface where the nurse will be working should be cleaned with a 20-sec alcohol friction rub, and covered with a sterile towel.

All principles of aseptic technique must be followed. The nurse should mask and wash his/her hands thoroughly with a povidone-iodine scrub.

A medication injection site is inserted into a closed port of the plastic bag. Each additive (electrolyte mix, trace elements, vitamins) is drawn up in a separate syringe and added one at a time through the cleaned medication injection site. Following each addition, the solution is thoroughly mixed by gentle agitation and the bag inspected for precipitants, particulate matter, or any solution change. If this is noted, the pharmacy is contacted and the bag discarded.

This method is utilized for several reasons. Since the patient load at the Toronto General Hospital is quite high, the pharmacy does not have the facilities, personnel, or time to make solutions complete with additives — except in special cases. In addition, there is very little waste in supplies and time. The additive that is changed most frequently is the electrolyte content. The patient's intake-output, weight, and laboratory values are monitored daily and, depending on the estimated electrolyte needs, the additive content of the solution is changed. For this purpose, a written order is completed by the physician and then sent to the pharmacy. The nurse, as a member of the parenteral nutrition team, is also alerted and since the addition is made by the nurse the change can be instituted immediately.

The nurse must receive additional training on correct mixing procedures and then follow them exactly. It does require additional nursing time but with continual practice the nurse becomes adept and efficient at the procedure.

In conclusion, this method is economical in both time and money and gives the nurse a greater feeling of satisfaction because of this increased responsibility. It may be added that using this technique in the ten-year history of the program at the Toronto General Hospital there has never been a septic episode due to contaminated parenteral nutrition fluid.

Once the additive solution is hung, it should not be left up longer than 12 hr. This rule ensures that the vitamins being administered are not degraded, and prevents the solutions from becoming contaminated by possible bacterial proliferation at room temperature.

Special additional additives such as sodium bicarbonate, cimetidine, insulin, etc. must be checked with the pharmacy. Some additives may shorten the life of the solution or the additive dose may be lower when given with the TPN regimen. There is always a possibility of component interaction and microprecipitation and careful inspection of the final solution should be done before and during its administration. The details of additive interaction and compatibility are given in Chapter 17.

Longevity of the catheter can be increased by refraining from using it to infuse a number of other solutions and/or medications. When additional supplementation is necessary it should be given through a separate intravenous line, i.e., a peripheral vein. If an alternative site is not available, then the total parenteral nutrition line may be used.

Although the author and associates have had success with this method of dividing responsibility between nurse and pharmacy, it must be stressed that this has been possible because of careful attention to aseptic technique by trained nurses who adhere to strict protocol as outlined.

V. PREPARATION OF THE FAT EMULSION

There are several fat emulsions now available on the market — all of them free of any of the significant side effects observed with fat emulsions available in the past. During the last ten years, considerable research and experience in intravenous nutritional support with fat emulsions has been undertaken.

When intravenous fat emulsions are given in conjunction with carbohydrates, they supply the required daily amount of energy as well as essential fatty acids which cannot be synthesized by the body. The products available are a soybean or a safflower oil-egg phospholipid emulsion which has the same properties as chylomicrons and is well tolerated in animals and man.

The fat emulsion should be kept refrigerated at 4°C until an hour before use. It then is allowed to reach room temperature to prevent any vasoconstriction when infused. Recently fat emulsions have been marketed that do not need refrigeration.

The nurse should aseptically remove the metal cap protecting the rubber diaphragm through which the giving set is to be inserted. The diaphragm should be cleaned with alcohol and, if desired, heparin is injected and the bottle is shaken gently to ensure complete mixing.

It is believed that the rate of removal of the intravenous fat emulsion is enhanced by the administration of heparin. The heparin releases or activates the clearing factor lipoprotein lipase (an enzyme) which hydrolyzes triglycerides to free fatty acids and glycerol. However, it has not been proven that the utilization of the fat emulsion is in any way enhanced by the addition of heparin. It should be remembered that the amount of heparin needed to induce this enzyme is very small, i.e., 1 unit/mℓ. The heparin should be added just prior to administration as its stability in fat emulsions is not known.

VI. ADMINISTRATION OF THE INFUSION THROUGH A CENTRAL VEIN

The technique for the administration of parenteral nutrition solutions via the central vein depends upon the nature of the infusate and also on a decision as to whether pumps are to be used.

When dextrose alone is the only source of nonprotein calories a pump is necessary in order to carefully control the inflow of hypertonic dextrose. In contrast, with a lower concentration of dextrose and use of substantial amounts of lipid as a calorie source, the need for pumps is less obvious and pumps may complicate the infusion regimen (by a need for two of them). In any case, the important aspect is that the parenteral nutrition solution be infused at a steady rate. Large or rapid changes in the infusion rate can result in significant metabolic complications, i.e., hypo- or hyperglycemia, caused by radical alteration in the rate of glucose administration. Whichever system is being used, it is important to check the flow rate frequently, i.e., every half hour. If the solution runs behind or ahead of schedule, the drip rate should not be accelerated or slowed down to meet the 24-hr volume requirements. Instead the drip rate should be adjusted to the correct hourly rate and continued at that rate.

At the Toronto General Hospital pumps are not used and staffers have found that the gravity drip method is quite suitable, especially for ambulatory patients. In fact all patients are encouraged to be as mobile as possible, which is easier when the patient does not have to push a pole with a cumbersome pump on it.

By avoiding the use of pumps we are able to keep costs at a minimum and do not have the worry of ensuring that all pumps are maintained in good working order. The nurses do not require additional in-service training on how to run the pumps and do not go through the process of "tricking" the alarms on the pumps. There is no chance of tubing joints being "blown apart", which can be a common occurrence with pumps.

If two solutions are to be infused simultaneously such as a glucose-lipid dual energy infusion system, two pumps would be required, thus making the situation rather complex for the nurse and frightening for the patient.

If a pump is preferred then it should be recognized that there are positive-pressure infusion pumps, peristaltic pumps, volume control pumps, etc., and the one selected depends on personal preference. If the pump is in good working order a constant rate of infusion should be observed. The correct tubing must be used with each pump, and with good connections to prevent joint rupture.

If pumps are used, usually only a simple amino acid-dextrose mixture is being administered with glucose as the main source of nonprotein energy. In this case commencement of therapy should be gradual to enable the body to increase insulin production to deal with the increased dextrose load. By infusing 1 ℓ the first day and then adding 1 ℓ a day based on each patient's metabolic response and caloric requirements, the final volume can be reached (2400 to 3000 mℓ/day). The higher the dextrose concentration, the slower the initiation and progression. This same rule applies to discontinuation of the total parenteral nutrition. The flow of the solution should be gradually decreased until the patient has adjusted to the lower rate of infusion. With such a system, any fat emulsion is usually given into a peripheral vein to provide essential fatty acids only at periodic intervals. Although conventionally 500 mℓ of 10% lipid emulsion is given only twice a week, the author recommends one bottle (500 mℓ) of 10% fat emulsion a day to ensure that the essential fatty acid content of phospholipid is within the normal range. When both fat and an amino acid-dextrose mixture are simultaneously infused into a central vein, the following regimen is used. The administration set used should be designed to assure a "closed system", preventing possibilities of contamination. The tubing system presently in use at Toronto General Hospital was developed to allow the infusion of various nutrients simultaneously, i.e., fat emulsion with the amino acid-dextrose mixture. This tubing has increased patient safety, reduced costs, and saved nursing time. The set consists of:

1. Tubing for the amino acid-dextrose mixture.
2. In-line filter for the above (0.45 or 0.22 μm).
3. Vented tubing for the fat emulsion.
4. Y-connector so the solutions can mix and be infused concurrently (supplied by Pharmacia Ltd., Montreal, Canada).

The above is also available preconnected in one package. Any connections are Luer-locked and when tightened correctly remain intact without reinforcing tape. This closed system avoids the need for any "needle piggybacking" or stopcocks which increase the risk of contamination.

The tubing is normally changed every 48 hr and the filter every 24 hr at 0700 hr in the morning, when the new solution containers are hung. The changing of the containers and tubing is performed by members of the nursing staff. All have received in-service education and supervision to ensure that all principles and proper techniques are obeyed.

In addition to the above set, extension tubing (40 cm in length) is attached to the catheter hub, so that there is little manipulation of the central venous catheter. This tubing is changed every 7 days with the new administration set at 0700 hr. This procedure should be done with the patient lying flat and performing the Valsalva maneuver, to prevent air embolism. In the ten years the Toronto General Hospital has had a TPN program, staffers have found the use of this additional tubing very beneficial. The patient is exposed to less risk at tubing change, catheters have less chance of being dislodged, cracked, or infected, and the nurse saves time and energy. It is much more comfortable for the patient, and by coiling this tubing once or twice on the chest and securing it with tape, there is no danger of the catheter being pulled out of its intravenous site. This coil absorbs the shock of any tug to prevent it being transmitted to the catheter.

As stated earlier in this text, the catheter should not be used for the infusion of other solutions and medications. This applies especially to blood products, which have a tendency to adhere to plastic catheters, thus becoming a prime target for the possible collection of bacteria. However, sometimes it is impossible to obtain an alternative route and the total parenteral nutrition line must be used. If this is the case, the following combination of infusions may be given:

1. Use of replacement fluid (normal saline) to run concurrently or intermittently with the parenteral nutrition.
2. Compatible intravenous medications (i.e., cimetidine) can be given into an additional fluid line while the parenteral nutrition infusion is in progress.
3. Incompatible medications (i.e., digoxin, antibiotics) require that the parenteral nutrition infusion must be stopped, the tubing flushed with a neutral solution, the medication instilled in the appropriate medication injection site and given over the prescribed period of time, the tubing flushed again, and the parenteral nutrition resumed.

To meet the needs of the above situations the parenteral nutrition administration set should lend itself to the attachment of a supplementary intravenous set or burette.

Another question is that of in-line filters. At present, the need for the use of in-line filters on the amino acid-dextrose side is a controversial issue. The Toronto General Hospital staff uses a 0.45 μm filter and feel it is advantageous since the nurses on the ward instill additives. This filter removes particulate matter and most bacteria, with the exception of *Pseudomonas*, and works well with a gravity drip. A 0.22 μm filter removes all bacteria and fungi, but may need a pump for the maintenance of proper flow rates. The use of a filter depends upon the mixing facilities, the knowledge of the staff using them, hospital policy, and budget.

Chapter 10

PARENTERAL NUTRITION THROUGH A PERIPHERAL VEIN

Gail Kennedy

TABLE OF CONTENTS

I. INTRODUCTION

Much attention has been given to the use of parenteral nutrition via a central vein, and not enough to a system that can be infused peripherally. This section will deal with the method which uses peripheral veins to infuse the TPN mixture. Details of catheter insertion and maintenance and techniques of infusion will be given in this chapter.

Peripheral parenteral nutrition is easy to undertake and can be managed by nurses or intravenous teams. This system can accommodate the needs of a wide range of medical and surgical patients whose caloric requirements are not exceedingly high and who have good peripheral veins.

The duration of treatment depends on the availability of veins in the individual patient, but it does allow us to treat patients who otherwise may receive routine intravenous therapy not meeting their total caloric and nitrogen requirements. Patients receiving parenteral nutrition by this method may fall into one of the following categories:

- Need for intravenous nutrition for a short period (10 to 14 days)
- Patients with a predisposition to septicemia or bleeding
- Patients with traumatized central veins
- Prospective long-term parenteral nutrition patients awaiting the insertion of a permanent catheter
- When central venous catheterization may be dangerous and inadvisable, i.e., radical head, neck, and chest surgery
- Patients with malnutrition, capable of eating but absorbing only suboptimal calories

A major consideration is still the fact that the site of peripheral vein infusion is a target for bacterial invasion. Thus, formal protocols must be developed to guide nurses and other related personnel in insertion and maintenance of such an infusion site so as to reduce the potential for infusion-related complications.

II. PERIPHERAL CATHETER INSERTION

To ensure a safe, efficient routine acceptable to the patient, he or she must be carefully prepared with a detailed explanation of the infusion composition, purpose, and techniques of initiating and maintaining this therapy.

The catheter should be inserted by a qualified nurse or an experienced doctor. In order to quickly recognize and correct any local complications only superficial peripheral veins are utilized. All principles of intravenous technique must be adhered to:

1. Select the appropriate vein taking into consideration its location and condition, the purpose of the infusion, and duration of therapy.
2. Select an appropriate needle.
3. Properly prepare the venipuncture site.
4. Use good insertion technique.
5. Dress the insertion site.
6. Observe for complications.

After the patient has been prepared and all questions answered, the necessary equipment should be assembled at the bedside. Equipment routinely used includes (Figure 1):

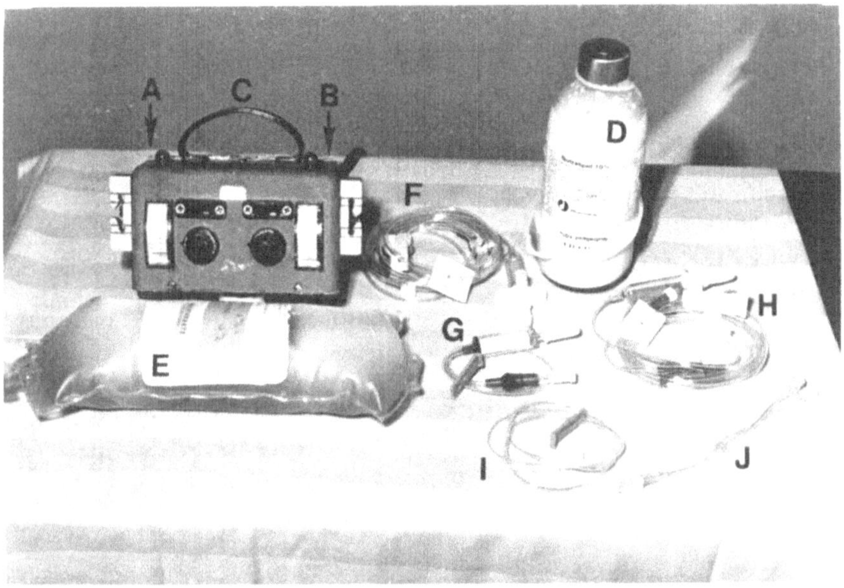

FIGURE 1. Dual-channel Holter® pump, solutions, and tubing for peripheral infusion. (A) Channel 1; (B) Channel 2; (C) pump; (D) lipid; (E) amino acid-dextrose mixture; (F) tubing for amino acid-dextrose mixture; (G) filter; (H) lipid tubing; (I) extension tubing; (J) Y-connector.

- 18-, 20-, or 21-gauge Teflon® catheter (e.g., 2″ Deseret®, Angiocath®)
- Alcohol swabs
- Povidone-iodine ointment
- Small gauze pad (½″ × ½″)
- Gauze (4″ × 4″)
- Op-Site® (5 cm × 7.5 cm)
- Nonallergenic tape
- Tourniquet
- Masks
- Prepared parenteral nutrition solutions
- Prepared tubing (flushed with infusate)
- Pump
- Intravenous pole

As mentioned, the insertion procedure begins with the selection of a good peripheral vein, ideally in the hand or arm. The tourniquet is applied, the chosen site is shaved and defatted, and is then followed by a liberal application of povidone-iodine solution. This is left on the skin for 2 min and the excess is removed with a sterile gauze. The Teflon® catheter is inserted and positioned so as to ensure a good blood flow. Then the parenteral nutrition tubing is connected, the tourniquet released, and the flow rate established with a pump or by gravity. After the tubing has been secured, the catheter site dressing is applied.

III. CATHETER MAINTENANCE

The catheter insertion site is covered with the gauze pad to which povidone-iodine ointment was applied. Application of Op-Site®, a small, transparent, air-occlusive dressing, further secures the catheter and protects the site. The dressing should not be so large as to interfere with close monitoring of the insertion site for signs of inflammation, swelling, etc. By rotating the site and changing the dressing every 48 hr, many complications can be avoided.

Safe and successful therapy depends upon the knowledge and skill of the nurse caring for the patient. Adherence to sterile technique is very important to prevent infection. The nurse should inspect the catheter site every hour (at the time of checking the drip rate) so that infiltration and inflammation are recognized. Failure to recognize these problems could result in damage to the surrounding tissues, limit the available veins for future therapy, and cause the patient not to receive his or her necessary nutrition. In case of extravasation of the infusion, the catheter should be removed immediately and the infusion restarted in a different vein. Some patients may be instructed to recognize these signs and immediately report them to the nurse.

Intravenous therapy exposes the patient to numerous hazards, many of which can be avoided if the nurse understands the risks and knows how to prevent their occurrence. Local complications occur most frequently while systemic complications are less frequent, though serious.

IV. COMPLICATIONS

The following are some of the complications related to peripheral parenteral nutrition catheters.

A. Thrombophlebitis

The vein receiving the infusion becomes very inflamed and a thrombus develops. If the infusion is allowed to continue, the vein becomes hard, tortuous, and very painful. Attention to technique can prevent many of these complications. The most important principle is to only infuse a mixture of equal volumes of fat and the amino acid-dextrose mixture. At no time should the latter be infused alone, either by design or inadvertently, due to improper flow through the drip. The author recommends a dual channel pump to ensure even flow. Early detection will prevent this complication for which the catheter should be removed.

B. Infiltration

If the catheter is dislodged, the solutions will be infused into the surrounding tissue. If this is allowed to continue, severe edema and necrosis could occur. Again, early detection will prevent this complication. If it does happen the catheter must be removed immediately.

C. Embolism

An embolus could occur from air, fat, or a free-floating blood clot. The symptoms associated with this are decreased blood pressure, weak rapid pulse, and cyanosis. If these are noted, the patient should be turned to his or her left side and put in the Trendelenburg position and the physician notified immediately.

D. Pyrogenic Reaction

This occurs when pyrogens are introduced into the bloodstream, producing a febrile reaction. Special precautions can reduce the incidence of pyrogenic reactions, such as the use of pyrogen-free solutions and protection of the solution from contamination. If a temperature elevation occurs, the parenteral nutrition should be halted and a protocol comparable to that used with central venous infusions followed (see Chapter 6).

V. INFUSATE COMPOSITION

The peripheral infusate will supply protein, calories, trace elements, and vitamins. However, the main problem in infusing TPN into a peripheral vein is the osmolality of the infusate. Very hypertonic solutions cause venous thrombosis and hence the major part of calories has to be provided by an isotonic substrate, namely lipid. In addition, there is some evidence that lipid emulsions may protect the vein. Because of these two considerations, it is vital that lipid be infused continuously and concurrently with an amino acid-dextrose mixture of a relatively low osmolality, using a Y connector as indicated below.

A commercially available solution containing 8 to 10% amino acids is mixed with an equal volume of 25% dextrose for a final concentration of 5% amino acids and 12.5% glucose, respectively. This mixture is infused with an equal volume of 10% fat emulsion. By infusing 1500 mℓ of the above amino acid-dextrose mixture and 1500 mℓ of lipid, approximately 2400 kcal can be given in a volume of 3ℓ/24 hr. The final mixture below the Y connector is slightly hypertonic (600 mOsm), but is well tolerated by most patients.

VI. ADMINISTRATION OF THE PERIPHERAL INFUSION

This system can be administered by gravity drip as safely and efficiently as any other peripheral intravenous system. Of course, the nurse must carefully monitor the infusion to ensure a constant rate of flow, preferably hourly.

At Toronto General Hospital the staff have chosen to aid the administration of peripheral parenteral nutrition with a pump. The pump of choice is the Holter® pump, which is a two-channel peristaltic infusion pump. It incorporates a variable speed motor which turns the rotor assembly. Pumping action is achieved by stretching and compressing the silicone rubber pump tubing as the rotor assembly turns. This pump ensures that the required proportion of lipid to amino acid-dextrose is delivered at the correct rate. By keeping this proportion constant the osmolality of the infused solution (less than 600 mOsm) will remain within acceptable limits and will not irritate the vein. It cannot be emphasized enough that the lipid emulsion must mix with the amino acid-dextrose solution at all times to ensure vein protection. This is achieved through the use of an infusion set with a Y connector, which allows a common segment for the mixing of the two liquids prior to entering the catheter.

If a pump is used, a thorough understanding of the operating instructions and maintenance requirements by all nursing personnel will assure maximum trouble-free operation. The flow rate and contents of the container should still be carefully monitored. Strict aseptic technique must be observed when handling the administration set. The pump should be kept clean, and any pump malfunction should be noted and repaired.

Again, if additional intravenous fluid or medication is required, an alternate site should be used. If this is impossible, then the parenteral nutrition line may be utilized. The same guidelines for giving additional fluids or medications must be adhered to as those for the central venous method (see Chapter 9).

READING LIST
(Chapters 5 to 10)

Bernard, R. and Stahl, W., Subclavian vein catheterization: a prospective study: non-infectious complications, *Ann. Surg.*, 173, 184, 1971.

Borresen, H., Coran, A. M., and Knutrud, O. L., Metabolic results of parenteral feeding in neonatal surgery: a balanced parenteral feeding program on a synthetic L-amino acid solution and a commercial fat emulsion, *Ann. Surg.*, 172, 291, 1970.

Brennan, M., O'Connell, R., Rosol, J., and Kundsin, R., The growth of *Candida albicans* in nutritive solutions given parenterally, *Arch. Surg.*, 103, 705, 1971.

Coran, D. and Nesbakken, R., The metabolism of intravenously administered fat in adult and newborn dogs, *Surgery*, 66, 922, 1969.

Deeb, E. N. and Natsios, G., Contamination of intravenous fluids by bacteria and fungi during preparation and administration, *Am. J. Hosp. Pharm.*, 28, 764, 1971.

Fischer, J., *Total Parenteral Nutrition*, Little, Brown, Boston, 1976.

Grant, J., *Handbook of Total Parenteral Nutrition*, W. B. Saunders, Philadelphia, 1980.

Hoshal, V., Total intravenous nutrition with peripherally inserted silicone elastomer central venous catheters, *Arch. Surg.*, 110, 644, 1975.

Hughes, E., Collective review. Venous obstruction in upper extremity, *Intern. Abstr. Surg.*, 88, 89, 1949.

MacDonald, A., Master, S., and Moffitt, E., A comparative study of peripherally inserted silicone catheters for parenteral nutrition, *Can. Anaesth. Soc. J.*, 24, 263, 1979.

Norden, C., Application of antibiotic ointment to the site of venous catheterization — a controlled trial, *J. Infect. Dis.*, 120, 611, 1969.

Qureshi, G. and Lilly, E., Complications of CVP catheter insertion in cubital vein, *JAMA*, 209, 1906, 1969.

Rowlands, D., Wilkinson, W., and Yoshimura, N., Storage stability of mixed hyperalimentation solutions, *Am. J. Hosp. Pharm.*, 30, 436, 1973.

Shoulders, H., Meng, H., and Tuggle, S., The effects of heparin on body temperature and plasma lipids following intravenous administration of fat emulsion in man, *J. Lab. Clin. Med.*, 52, 559, 1958.

Vanderwyk, R., Microbicidal Effectiveness of Betadine Surgical Scrub Versus Hexachlorophene, *Purdue-Frederick, Yonkers, N.Y., 1971.*

Walter, M. D., Stanger, H., and Rotem, C. M., Complications with percutaneous central venous catheters, *JAMA*, 220, 1455, 1972.

Wilson, J., Grow, J., Demong, C., Prevedel, A., and Owens, J., Central venous pressure in optimal blood volume maintenance, *Arch. Surg.*, 85, 55, 1962.

Zinner, S., Denny-Brown, B., Braun, P., Burke, J., and Kass, E., Risk of infection with intravenous indwelling catheters: Effect of application of antibiotic ointment, *J. Infect. Dis.*, 120, 616, 1969.

Chapter 11

HOME TOTAL PARENTERAL NUTRITION

Gail Kennedy and Khursheed N. Jeejeebhoy

TABLE OF CONTENTS

I. INTRODUCTION

Total parenteral nutrition (TPN) is the technique of supplying all necessary nutrients (protein, fat, carbohydrate, vitamins, minerals, and trace elements) intravenously to meet the requirements of patients when gastrointestinal (GI) diseases prevent the ingestion of an adequate oral diet. The development of this modality has given us time for the resolution of a disease without the complication of malnutrition.

TPN has been successfully employed in the hospital setting and has now been extended to the ambulatory patient requiring "artificial" nourishment for a prolonged period of time. The development of this now-recognized modality has allowed patients to become independent of the hospital and active in society.

In the following sections, physicians, nurses, pharmacists and patients in the program will find all the details necessary to institute and maintain the program in an easy and safe manner.

II. THE TPN TEAM

A. The Team Approach

It is a difficult task to accurately assess whether a patient is a good candidate for the home TPN program. Initially, the patient should be assessed for his/her ability (1) to learn, (2) to handle the technical aspects of the infusion system, and (3) to emotionally and psychologically cope with the system. In addition, the availability of a family to provide acceptance, encouragement, and support are essential to a successful outcome. Furthermore, the patient should have access to a team of experts who can easily be contacted, to encourage, support, and meet any emergencies. Such a team is composed of:

Physician
Registered nurses
Pharmacists and technicians
Social worker
Psychiatrist

These team members should be available to encourage the patient to verbalize concerns, beliefs, fears, and feelings about the system.

Prior to the insertion of a permanent catheter, it is a good time to answer questions, provide information, offer support, and even start a part of the training program. This same team should continue to be involved with the patient at home, following discharge from the hospital.

B. Teaching Facilities

The new home TPN candidate is transferred to a designated nursing unit where the staff is proficient in home TPN skills. Here the patient is taught the technical skills, based on the expertise developed by the TPN teacher, head nurses, and staff nurses of that unit. Although other forms of organization may be used to teach a patient, the above concept provides continuity of care and personnel. This is especially useful when the patient initially may have come to the hospital acutely ill and have been brought from that state to health by that staff. As well, these nurses are often involved in other aspects of patient care and thus can look at the patient's needs as a whole.

C. Role of the TPN Teacher*

1. To develop the overall teaching program for home TPN.
2. To develop and update a reference manual containing guidelines for nurses and home patients in the programs.
3. To act as a resource person capable of clarifying the principles and methods of teaching and evaluation of results.
4. To periodically evaluate the effectiveness of the nurse's teaching and the patient's progress.
5. To continually examine techniques and guidelines to improve the standards of nursing care in the light of new knowledge.
6. To plan and coordinate the care of these patients with the total team.
7. To act as a liaison between the patient, family, TPN team, and community health resources.
8. To provide support, encouragement, and answers to problems arising after discharge by follow-up contact (visits, telephone).
9. To assist in educating the nurses in the community about the TPN system.

D. Role of the Head Nurse**

1. To instruct, maintain, and standardize the nursing staff's knowledge and skills about home TPN.
2. To monitor the home patient during the teaching period for the purpose of assessing the effectiveness of the teaching.
3. To evaluate the patient's knowledge and technique prior to discharge.
4. To participate in the TPN team promoting continuity and quality of care.

* One person from the nursing staff should assume the role of the teacher. The person and his/her departmental affiliation will depend on local conditions at the institution.

** These remarks apply to a system where the patient is taught in the nursing areas of the hospital. In some institutions this is all done in pharmacy. The Toronto General Hospital prefers the former approach.

E. Role of the Staff Nurse*

The staff nurses are responsible for teaching the patient about the system, under the supervision of the head nurse and TPN teacher. The head nurse should attempt to assign nursing staff on a consistent basis for the purpose of teaching. With careful planning, one nurse could teach the patient for at least three days in succession.

The nurse conducting the teaching sessions must (1) be proficient in the knowledge of home TPN and (2) demonstrate an understanding of the principles of teaching. The nurse must appear calm, confident, and relaxed in order to achieve maximum benefit in the session.

When developing the training program the following teaching and learning principles are useful to consider.

- Learning is more effective when in response to a need felt by the learner. The teaching program can only be effective if the patient is ready to accept the idea of home TPN. Severe depression, denial, or anger should be dealt with prior to starting the patient in the program.
- New learning must be based on previous knowledge and experience. A good teacher will assess the patient's background knowledge of medical terminology and skills. In addition, assessment of the patient's physical and emotional status prior to the session will result in the process becoming more productive and enjoyable for both teacher and student. Encourage the patient to keep a workbook with a summary of points to be remembered. Then, prior to repeating the procedure, these notes can be reviewed.
- An individualized program promotes retention and application of learning. Each patient should be considered individually in regard to the approach and the rate of teaching. The stages of progressive teaching consist of allowing the patient to first view the procedure, and gradually assuming increasing responsibility until total responsibility is assumed by the patient. No attempt should be made to push the patient, as this may cause errors which in turn lead to feelings of frustration and discouragement. A minimum number of people should be involved in the teaching session so as to allow for maximum concentration and minimum anxiety. Communication is the key point — share observations about progress and encourage feedback from the patient.
- People learn by doing, therefore active participation on the part of the learner is essential. Early involvement is desirable. As soon as the patient feels comfortable with any aspect of the procedure, it should be practiced. Save all criticisms to the end and present them as positive alternatives. Correcting a patient must be done with tact — interrupt only as a last resort; for example, in the case of a dangerous error such as contamination. Interruptions may induce anxiety and frustration and lead to increased errors.
- Meaningful repetition reinforces learning. The nurse's charting is a valuable tool for the ongoing evaluation of the progress of the patient's teaching. Accurate notes ensure that weak areas are recognized and corrected. Notes will also enhance consistency in the approach to the training. Each note should end with a brief list of the areas to be emphasized or reviewed the following day.
- Moving from simple to complex ideas gives the learner a sense of achievement. The program outlined below is designed on this principle. In the following sections it will be evident that the trainee starts off learning simple things, such as handwashing and sterile technique, and ends by actually connecting the infusion system and running the solutions.

* These remarks apply to a system where the patient is taught in the nursing areas of the hospital. In some institutions this is all done in pharmacy. The Toronto General Hospital prefers the former approach.

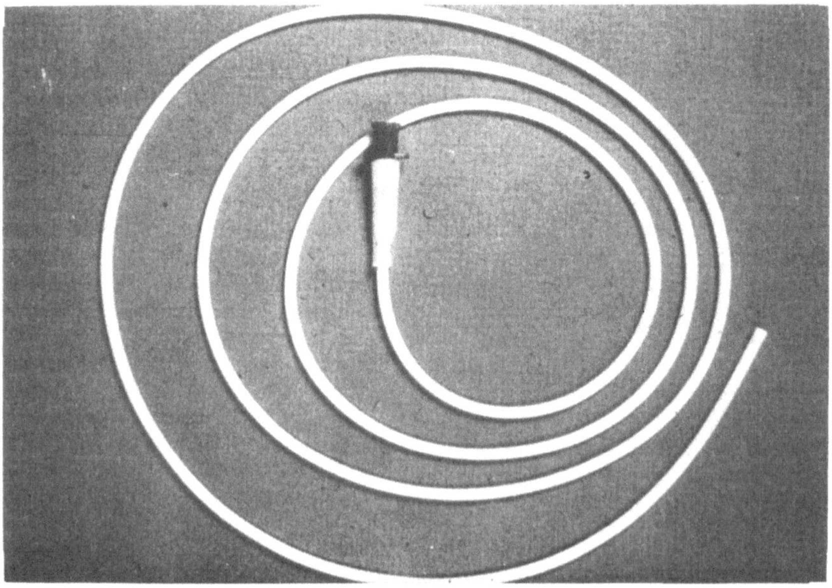

FIGURE 1. Silastic® (silicone rubber) "Langer" catheter is 100 cm long (I.D. 0.76 and O.D. 2.72 mm) but cut to appropriate length at time of insertion to allow it to lie in vena cava just outside right atrium. Note permanent plastic Luer-lock hub.

- The environment can be used to focus the learner's attention and avoid distraction. A teaching area where privacy promotes concentration and an exchange of ideas and feelings is essential. Ideally, the room should contain a sink, a stove, a mirror and other necessary equipment to simulate the home environment. It is believed that a smoother transition from hospital to home will then occur.
- Incentives motivate the learner. Always end each teaching session on a positive note. Summarize the areas performed well.

III. PLACEMENT AND CARE OF THE PERMANENT CATHETER FOR TPN

It is now generally agreed that a silicone rubber catheter placed in the superior or inferior vena cava via the subclavian, jugular, or external iliac veins constitutes the most satisfactory vascular access system. The intravenous solutions are readily diluted here and these large veins can tolerate the insertion of a long-term indwelling catheter.

A Silastic® tube (silicone rubber) (Langer catheter, Figure 1), 100 cm in length with an internal diameter of 0.76 mm and an external diameter of 2.72 mm is used. It is connected to a permanent plastic Luer-lock hub.*

The catheter is tunneled subcutaneously for at least 18 to 20 cm before entering one of the above-mentioned veins through a short side branch, such as the cephalic, common facial, or hypogastric vein, then threaded through to the subclavian (Figure 2A), jugular (Figure 2B), or external iliac vein (Figure 2C), respectively, to the vena cava under fluoroscopic control. It is tied into place using a loop-like stitch subcutaneously. Externally, a Teflon® piece is glued to the Silastic® catheter using tetrahydrofuran and is stitched to the skin to fix it temporarily. This reduces the risk of catheter migration.

* Available through Extracorporeal Medical Specialties, Inc., King of Prussia, Pa., U.S.A. Made to Toronto General Hospital specifications.

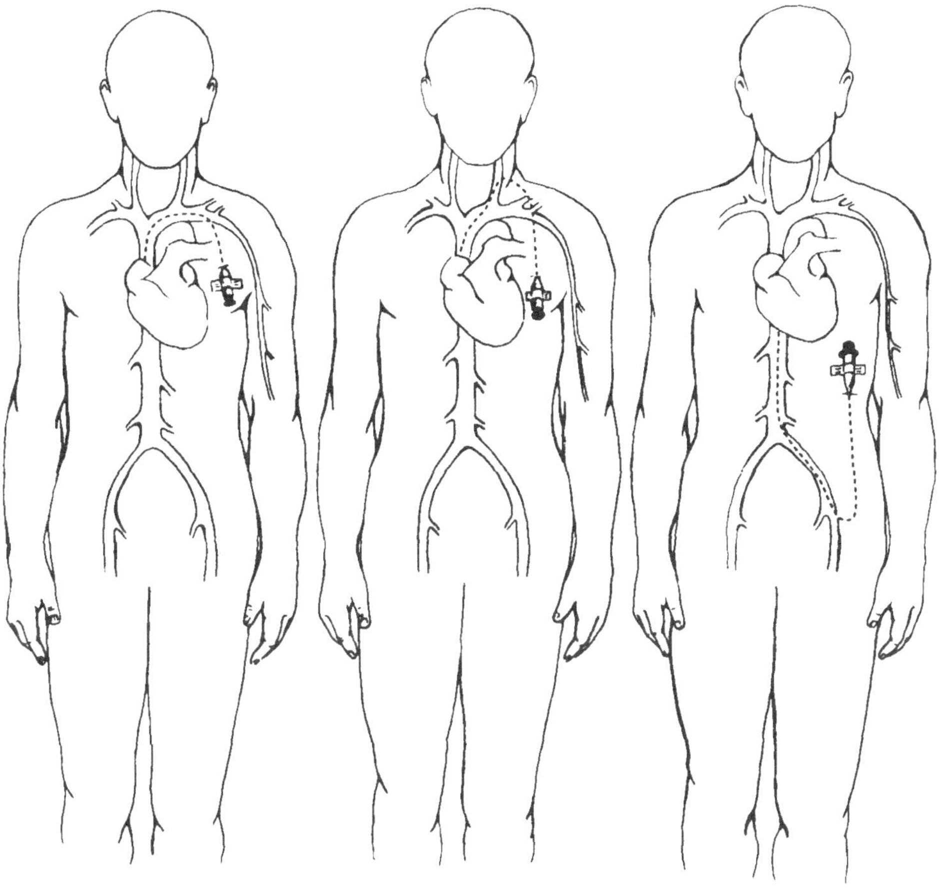

FIGURE 2. Sketch to show insertion sites for permanent vascular access by central vein catheter. These may be on the right or left side of torso but are shown only for the left side. The sites are for reaching the vena cava via (A) subclavian, (B) internal jugular, or (C) external iliac vein.

It is believed that the tunnel helps to minimize the risk of infection. It also makes it easier to handle the catheter and to care for the exit site. This entire procedure is done under a general anesthetic in the operating room.

TPN has one major hazard and that is the possibility of introducing systemic infection. This is especially so because of the long-term indwelling catheters, the infusion of nutrient solutions capable of supporting bacterial and fungal growth, or sources of infection in the form of existing fistulae, ileostomies, colostomies, G-tubes, etc. Aseptic technique is the key to success and the freedom from complications. Asepsis signifies the absence of organisms, and the patient should be so indoctrinated and encouraged to put this vital concept into practice through attention to the following matters.

A. Principles of Aseptic Technique

1. A sterile object becomes contaminated when in contact with an unsterile object, i.e., sterile forceps are no longer sterile if they touch an unsterile object, and can no longer be used until resterilized.
2. Never walk away from or turn away from a sterile field. This will prevent possible contamination while the field is out of view.

3. Hold sterile objects above the level of the waist. This will ensure the object is in sight, thus avoiding accidental contamination.
4. Avoid talking, coughing, sneezing, or reaching over a sterile field or object. This principle will help to prevent contamination by droplets from nose and mouth, or by particles from an arm or hair.
5. Liquids flow in the direction of gravity, i.e., keep forcep tips held down. If they point upwards, the disinfectant solution runs down and touches an unsterile part of the forcep, then when pointed downward again, it flows back to the sterile part and the forceps become contaminated.

Thorough handwashing is included as part of aseptic technique. Evidence in the literature supports the importance of careful and frequent handwashing. Contaminated hands are considered to be the prime factor in infections.

B. Handwashing Procedure

1. Transient bacteria, the type that accumulates on hands in everyday living and working, are easily removed by thorough washing with soap and water. Dial® or Zest® soap (contains hexachlorophene) should be used.
2. Turn on water and regulate the temperature. Leave the water running throughout the procedure.
3. Wet hands and forearms under running water.
4. Start at fingertips by cleaning nails with an orange-wood stick or a nail brush.
5. Apply soap over hands, wrists, and forearms. With hands in downward position, rub vigorously for approximately 2 min. Start at fingers, then over palms, the back of hands, wrists, and then halfway up forearm. Never return to an area just rubbed. Work from cleanest to dirtiest — fingers to forearms.
6. Rinse hands and arms in the same manner as above, under running water. Hold hands up after rinsing so water will not run down to hands.
7. Dry hands with paper towel or clean linen towel.
8. Turn taps off with paper towel, if possible.

After the patient understands asepsis and has mastered the handwashing technique, the dressing procedure may be taught.

C. Equipment and Sterilization

The patient is introduced to a simple sterilization technique applicable to the home setting. This makes him/her independent of prepackaged dressing trays and reduces costs. The equipment is sterilized by simply boiling it, as described.

1. Equipment

Large roasting pan	Dressing tray	
2 kitchen tongs (insulated handles)	1 metal container (detergicide)	
80% alcohol	1 or 2 thumb forceps	patient preference-total of 3
Kitchen tray	1 or Kelly forceps	
Mask	2 medicine glasses (povidone-iodine and alcohol)	

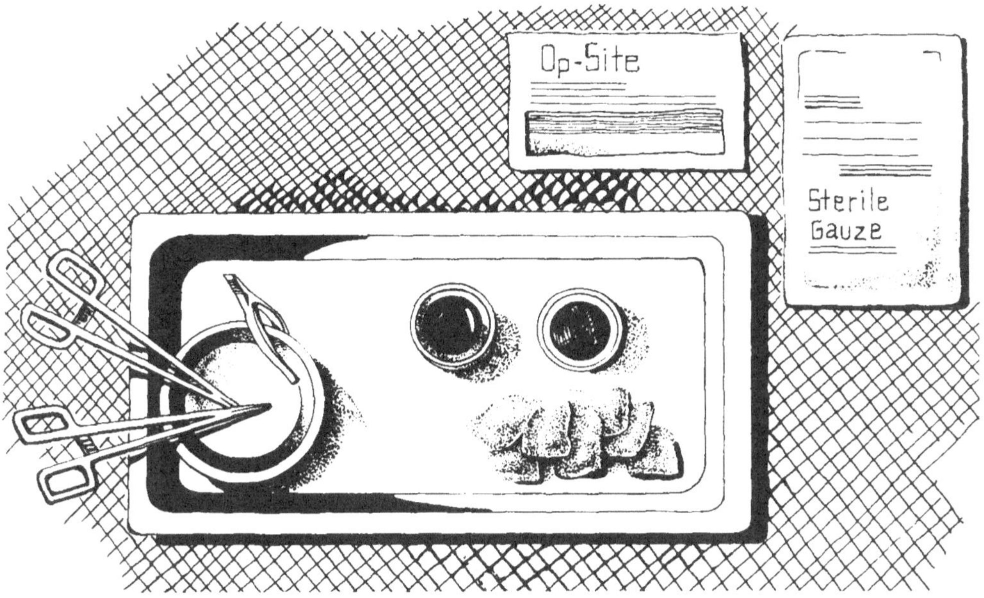

FIGURE 3. Sketch of dressing tray and contents as arranged on clean kitchen tray, ready for use, after sterilization at home. Note forcep and tweezers accessible for use without contamination. Large bowl may be used for detergicide and smaller ones for povidone-iodine and alcohol.

2. Sterilization Technique

1. Place all items of dressing tray in roasting pan. Fill the roasting pan with sufficient water so all items are covered.
2. Place the kitchen tongs on the edge of roasting pan so the tips are covered by water, but the handles are easily accessible.
3. Cover with lid.
4. Bring water to a vigorous boil. Allow the equipment to boil 15 min longer.
5. Clean kitchen tray with 80% alcohol using a 20-sec friction rub.
6. Mask.
7. Wash hands.
8. Using the tongs, carefully remove dressing tray from water, drain, and place it right side up on tray. Take out containers and forceps and place on sterile dressing tray. Place handles of forceps over the edge of the containers so they are easily accessible (Figure 3).
9. Carefully carry the kitchen tray holding the dressing tray to the area where the dressing will be done.

D. Cleansing of Catheter Exit Site

The point where the catheter leaves the skin tunnel is the exit site. This site must be carefully cleaned and dressed to prevent the growth of bacteria that could infect the catheter and tunnel.

There are two dressing procedures. The first, a gauze dressing (see Chapter 8, Figure 4) changed daily, allows for continual practice in dressing skills. When this dressing is performed satisfactorily, the Op-Site® dressing (Chapter 8, Figure 5) may be taught.

E. Technique for Gauze Dressing

1. Assemble

- Sterile dressing tray
- Detergicide 1:750

- 10% Povidone-iodine solution (P.I.)
- Povidone-iodine ointment (P.I.)
- Alcohol 80%
- 3 packages sterile cotton balls
- 1 package sterile 4″ × 4″ gauze
- 1 package sterile 2″ × 2″ gauze
- 1 package ½″ × ½″ gauze
- Nonallergenic waterproof tape
- 3″ strip of 1″ adhesive tape

2. Procedure for Gauze Dressing

1. A mask should be worn and hands washed well.
2. Bring sterile tray to area where dressing is to be done. It should be in front of a mirror to allow easy viewing of the site and catheter.
3. Open detergicide 1:750 bottle. Discard a small amount of this liquid to clean the lip of the bottle (pour-off technique) and then add the rest to the metal container.
4. Repeat above procedure with P.I. solution, into a medicine glass.
5. Repeat above procedure with alcohol, into the second medicine glass.
6. Carefully open packages of cotton balls and gauze. Do not touch the inside of the package. Allow the contents of the package to "fall" onto the dressing tray.
7. Separate the double layer of 2″ × 2″ gauze.
8. Open the tube of P.I. ointment. Discard a small amount. Add a dab to one layer of 2″ × 2″ gauze. Do not touch the gauze with any part of the tube.
9. Cut strips of the waterproof tape to the desired length and stick them to the edge of the table.
10. Remove the old dressing carefully (see Section III, G). Observe dressing for any discharge. Observe the exit site and surrounding skin for redness, swelling, or discharge. If noted, complete the dressing and inform the doctor.
11. If in the same room as the sink, wash hands again.
12. To check for catheter migration, measure the distance from the exit site to the catheter tab with a pair of marked forceps. A slight amount of catheter migration is expected over a period of time, but no more than 2″ should be allowed. If this distance has notably increased, complete the dressing and inform the doctor.
13. The skin is first cleaned with the detergicide 1:750. Start at the exit site and clean in a circular motion. Use a new cotton ball for each circle. Continue the circular motion to a distance totalling a 4″ diameter around the catheter.
14. With a cotton ball, start at the site and clean down the line, past the tab.
15. Clean the tab with a damp cotton ball.
16. Repeat numbers 13, 14, and 15 with the P.I. solution. Allow this to remain on the skin for 2 to 3 min.
17. Repeat numbers 13, 14, and 15 with alcohol.
18. Dry the skin with a sterile 4″ × 4″ gauze.
19. Using forceps, pick up the ½″ × ½″ gauze. Touch it to the ointment on the single layer of 2″ × 2″ gauze so there is a small amount on the ½″ × ½″ gauze. Apply this to the catheter exit site.
20. Using forceps, pick up the remaining 2″ × 2″ gauze and cover the exit site up to the tab.
21. Secure this with waterproof tape. If the tape does not adhere well, spray some tincture of benzoin onto the skin around the gauze. Allow this to dry and then apply tape. The tape should form an occlusive, waterproof dressing.

22. Remove the old adhesive tape and apply a new piece to secure the tab into position. Change the tape only two times per week unless it becomes soiled or is no longer attached to the skin.
23. Coil the extension tube away from the dressing and tape to chest. If tape becomes loose, retape as necessary.
24. This dressing is changed daily.

F. Technique for Op-Site® Dressing

Op-Site® is a hypoallergenic, transparent material that provides an occlusive, waterproof dressing. This dressing is only changed once per week unless (1) drainage is noted from the site or (2) it is no longer occlusive.

1. Assemble

- Sterile dressing tray
- Detergicide 1:750
- Povidone-iodine solution 10% (P.I.)
- Alcohol 80%
- 3 packages sterile cotton balls
- 1 package sterile 4″ × 4″ gauze
- 1 package sterile 2″ × 2″ gauze
- 1 package sterile ½″ × ½″ gauze
- Povidone-iodine ointment (P.I.)
- Op-Site® (5 cm × 7.5 cm)
- Mask

2. Procedure for Op-Site® Dressing

1. A mask should be worn and hands washed well.
2. Bring sterile tray to area where dressing is to be done. It should be in front of a mirror to allow easy viewing of the site and catheter.
3. Open detergicide 1:750 bottle. Discard a small amount of this liquid to clean the lip of the bottle (pour-off technique) and then add the rest to the metal container.
4. Repeat above procedure with P.I. solution, into a medicine glass.
5. Repeat above procedure with alcohol, into the second medicine glass.
6. Carefully open packages of cotton balls and gauze. Do not touch the inside of the package. Allow the contents of the package to "fall" onto the dressing tray.
7. Separate the double layer of 2″ × 2″ gauze.
8. Open the tube of P.I. ointment. Discard a small amount. Add a dab to one layer of 2″ × 2″ gauze. Do not touch the gauze with any part of the tube.
9. Remove the old dressing carefully (see Section III, G). Observe dressing for any discharge. Observe the exit site and surrounding skin for redness, swelling, or discharge. If noted, complete the dressing and inform the doctor.
10. If in the same room as the sink, wash hands again.
11. To check for catheter migration, measure the distance from the exit site to the catheter tab with a pair of marked forceps. A slight amount of catheter migration is expected over a period of time, but no more than 2″ should be allowed. If this distance has notably increased, complete the dressing and inform the doctor.
12. The skin is first cleaned with the detergicide 1:750. Start at the exit site and clean in a circular motion. Use a new cotton ball for each circle. Continue the circular motion to a distance totalling a 4″ diameter around the site.

13. With a cotton ball, start at the site and clean down the line, past the tab.
14. Clean the tab with a damp cotton ball.
15. Repeat numbers 12, 13, and 14 with the P.I. solution. Allow this to remain on the skin for 2 to 3 min.
16. Repeat numbers 12, 13, and 14 with alcohol.
17. Dry the skin with a sterile 4″ × 4″ gauze.
18. Using forceps, pick up the ½″ × ½″ gauze. Touch it to the ointment on the single layer of 2″ × 2″ gauze so there is a small amount on the ½″ × ½″ gauze. Apply this to the catheter exit site.
19. Carefully pull the paper backing off so ¾ of the Op-Site® is exposed. Do not touch this exposed area with your hands. Keep the remaining Op-Site® away from the skin.
20. Position the Op-Site® so it is above the tab and the catheter exit site then will be in the middle. Smooth this half down, especially around the catheter.
21. Ensure that the Op-Site® adheres around the line by pinching the Op-Site® together (to itself).
22. Gradually pull the backing off with one hand as you guide and smooth the remainder of the Op-Site® over the line.
23. Remove the old adhesive tape securing the tab and apply a new piece. Change when necessary.
24. It is not necessary to remove the green strip on the Op-Site® but it can be removed if desired. Carefully cut a mark where the green strip meets the clear part. Tear at this mark.
25. For males: if there is a lot of hair in the area covered by the dressing, it should be shaved. This ensures the Op-Site® remains in place and allows for easier removal. Soak a disposable razor in alcohol for 2 hr or boil with dressing equipment. After the soiled dressing is removed wet the skin with detergicide and shave the area from the exit site to the outer edge. Be careful not to nick the skin or cut the catheter. Continue with cleaning technique.

G. Removal of Dressing

1. With one hand, apply a slight pressure over catheter at exit site.
2. With the other hand, peel back one corner of the dressing to the line.
3. Go to each corner and repeat this.
4. Hold the line in place with a sterile forcep and remove the remainder of the dressing in an upward direction. This will prevent the catheter from being pulled out.
5. If it is difficult to remove the dressing (Op-Site®), use a small amount of adhesive remover on gauze and apply this to the lifted edge.

IV. COMPLICATIONS TO RECOGNIZE WHILE PERFORMING THE DRESSING PROCEDURE

A. Drainage

If redness or drainage is noted around the catheter exit site, inform the doctor for further instructions. Note the type, amount, and odor of discharge. A swab of the site (for culture for bacteria and fungi and sensitivity to antibiotics) may be needed. An antibiotic may be ordered until this problem is resolved. When drainage is observed, the Op-Site® dressing should be discontinued and a gauze dressing applied so as to be able to monitor the site daily.

B. Skin Reaction

A skin rash or irritation may develop from the waterproof tape or Op-Site®. This can be treated by applying tincture of benzoin to the irritated skin. Spray the benzoin on, allow it to dry, and then apply the tape. If the skin under the dressing continues to remain red, contact the hospital.

C. Premature Dressing Change

This is often a problem with the gauze dressing, for the waterproof tape sometimes does not adhere for more than 24 hr. Using tincture of benzoin may help the tape to adhere longer. Reinforce or replace as necessary.

D. Perspiration

In warm weather perspiration may increase and cause the Op-Site® to become loose and allow the accumulation of water. If this occurs, the dressing must be changed. Persistence of this problem, necessitating daily dressing changes, may require the use of a gauze dressing until the warm weather subsides.

E. Line Migration

The Teflon® tab should always be sutured to the skin to prevent the line from pulling out. Check the condition of the sutures during the dressing change. If they are loose or out, or if the line appears to have moved, contact the physician.

V. PREPARATION AND ADMINISTRATION OF INFUSIONS

Once proficiency is achieved with the dressing technique, the patient is taught to prepare and administer the infusion. The amino acid-dextrose mixture to which vitamins and additional intravenous drugs are added is usually infused during the night, over an 8- to 10-hr period while the patient sleeps. The lipid emulsion is infused separately. The order in which the amino acid-dextrose additive mixture and the lipid is infused should be the individual patient's choice, so as to be consistent with his/her lifestyle.

A. Preparation and Infusion of Lipid Emulsion

1. Assemble

- Bottle of lipid emulsion (500 mℓ)
- Tubing for infusing lipid (vented)
- Heparin (1000 units/mℓ)
- Alcohol swabs
- 1.0-mℓ Syringe
- 21-Gauge needle
- Adhesive tape (1″)
- Mask

*2. Procedure**

1. Remove the lipid emulsion from the refrigerator 1 hr prior to the infusion to allow it to reach room temperature.
2. Mask.
3. Wash hands.

* See Figure 4 to identify the parts and tubing to which reference is made in this section.

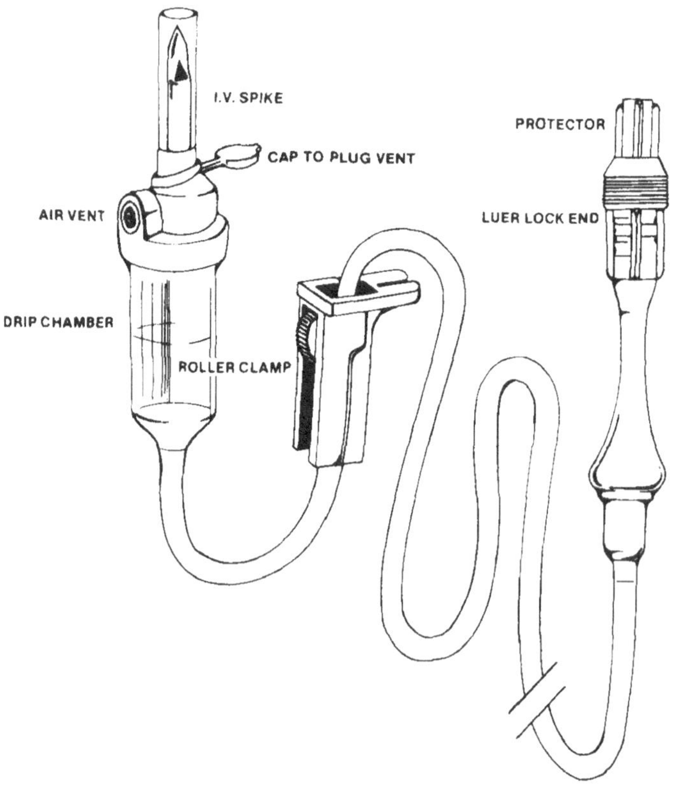

FIGURE 4. Sketch of tubing and component parts for connecting catheter with lipid emulsion. This is in a rigid bottle, thus requiring venting or the entry of sterile air to replace the emulsion as it flows out.

4. Remove the metal cap from the bottle and clean the rubber diaphragm with an alcohol swab.
5. Draw up 500 units (½ m*l*) of heparin. Inject this through the small cross on the rubber diaphragm of the bottle (Figure 5).
6. Gently agitate the bottle to insure the heparin has mixed with the lipid.
7. Remove the lipid tubing from its sterile package and close the roller clamp. Insert the spike into the large circle.
8. Gently squeeze the drip chamber until half full. Open the roller clamp and flush air out of the tubing.
9. With an alcohol swab, clean the connection between the extension tube and medication injection cap.
10. Remove the medication injection cap from the end of the primary extension tube. Pinch the end of the lipid tubing and extension tubing to raise the fluid level. Attach the lipid tubing.
11. Tighten connection and secure this with adhesive tape.
12. Open the slide clamp on the extension tube. Open the roller clamp on the lipid tubing and adjust the rate so that the lipid is infused over 2 hr (84 drips per min).
13. To maintain a constant rate since the lipid is infused by gravity, it is important to keep the height of the bottle higher than the catheter exit site.

B. Preparation and Infusion of Amino Acid-Dextrose Solutions

1. Assemble

- The required number of bags of solution
- Vitamin injection to be used as directed

FIGURE 5. Sketch of bottle of lipid emulsion showing sites in diaphragm (closure) for injection of additives (e.g., heparin) and for insertion of the delivery tube spike-end. The third site (small circle) is for insertion of a needle in those cases where a venting tube inside the bottle is supplied as an alternative to vented tubing for allowing air entry.

- Synkavite® — 10 mg/week
- Iron-dextran — 1 mℓ/day (diluted iron-dextran solution, 0.5 mg of iron/mℓ)
- Straight intravenous tubing
- IVEX-2® or other suitable filter set
- Medication injection site
- Plasma transfer set
- Extension tubing
- Syringes
- 25-Gauge needles
- Alcohol swabs
- Adhesive tape
- Mask

2. Procedure

1. Remove the required number of bags of the amino acid-dextrose solution from the refrigerator 1 hr prior to use. This will allow it to reach room temperature.

2. Assemble the above material in a quiet and clean working area. Clean the surface with 80% alcohol using a 20-sec friction rub.
3. Mask.
4. Wash hands.
5. Arrange the amino acid-dextrose solution bags on the working surface in the order in Figure 6.
6. Insert medication injection site with a gentle twisting motion into porthole 6. Keep the flaps of the portholes folded back and be careful not to touch the spike with the hand, or puncture the bag when inserting it (Figure 6).
7. Draw up the appropriate medications in the syringes. Change the needle on each syringe to a sterile 25-gauge needle.
8. Swab the medication injection site with alcohol.
9. Instill each medication through this site. Gently agitate the bag between each additive to ensure mixing of the medication.
10. Observe the bags for any precipitated matter. If there is any, discard the bag and take a fresh one. Inform the physician or pharmacy on the following day.
11. Remove the plasma transfer sets from the box. Close the roller clamps. Insert each spike into a port, i.e., one set is inserted into portholes 5 and 4, while the remaining set is inserted into 3 and 2 (Figure 6).
12. Remove the straight tubing from its box. Close roller clamp. Insert spike into porthole 1 (Figure 6).
13. Attach IVEX-2® filter set to the end of the straight tubing.
14. Attach an extension tube to end of IVEX-2® filter.
15. Place each liter of solution into its own 1-ℓ pressure infusion cuff (Figure 7).
16. Hang cuffs on intravenous pole.
17. Open roller clamps on plasma transfer sets.
18. Fill the drip chamber on the straight tubing to half by gently squeezing it. Remove protective cap from end of extension tube. Invert the filter and open roller clamp on straight intravenous tubing and flush out all the air. When air is out, close roller clamp and attach a sterile needle to the end of the extension tube to maintain its sterility.
19. When ready to "hook up", remove the needle from the end of the extension tube, pinch this tube and the primary extension tube to raise the fluid level and remove trapped air. Connect to the primary extension tube.
20. Tighten all tubing connections and secure with adhesive tape.
21. Turn on pneumatic infusion pump* (see Chapter 10, Section VI, for use of mechanical pump).
22. Regulate drip rate. The drip rate is determined by the volume of fluid to be infused over the prescribed period of time (8 to 10 hr). The time will vary with each individual and should be confirmed by the physician. For example: Mrs. A is to receive 3ℓ (3000 mℓ) over 10 hr. She will receive 300 mℓ in 1 hr. To determine the drops (gtt) per minute: 300 mℓ/hr/60 min = 5 mℓ/min 10 gtt = 1 mℓ 10 × 5 = 50 gtt/min.
23. Add 10 drops to the determined flow rate.
24. Establish the drip rate by using a watch and counting the number of drips in a minute.
25. One hour prior to termination of the infusion, decrease the flow rate to one half or one quarter of the original.

Toronto General Hospital HPN patients, given the choice of obtaining premixed bags of infusate or the option of mixing their own solutions, have favored the former.

* Alternately, a mechanical pump can be used. The manner of using such a pump will depend on the manufacturer.

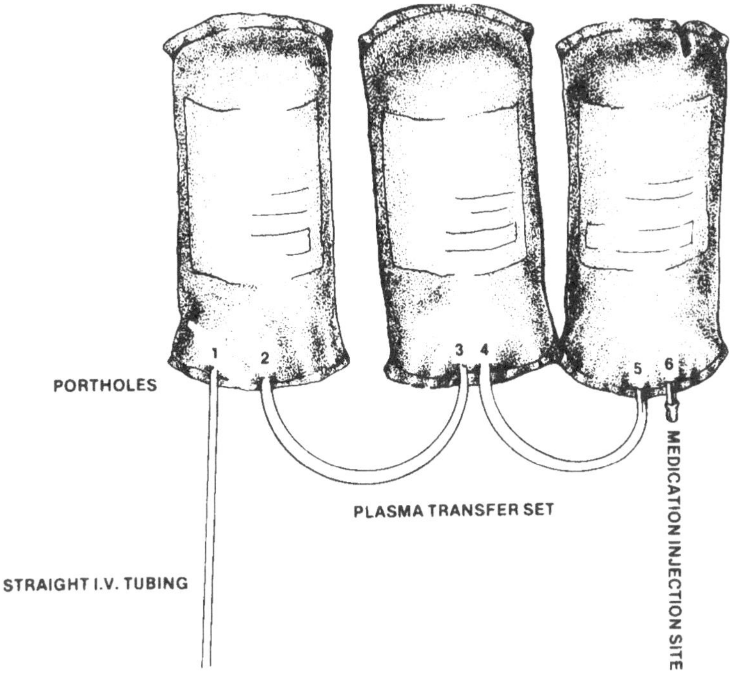

FIGURE 6. Sketch to show principles of connection of amino acid-dextrose bags for intravenous administration of their contents, usually overnight.

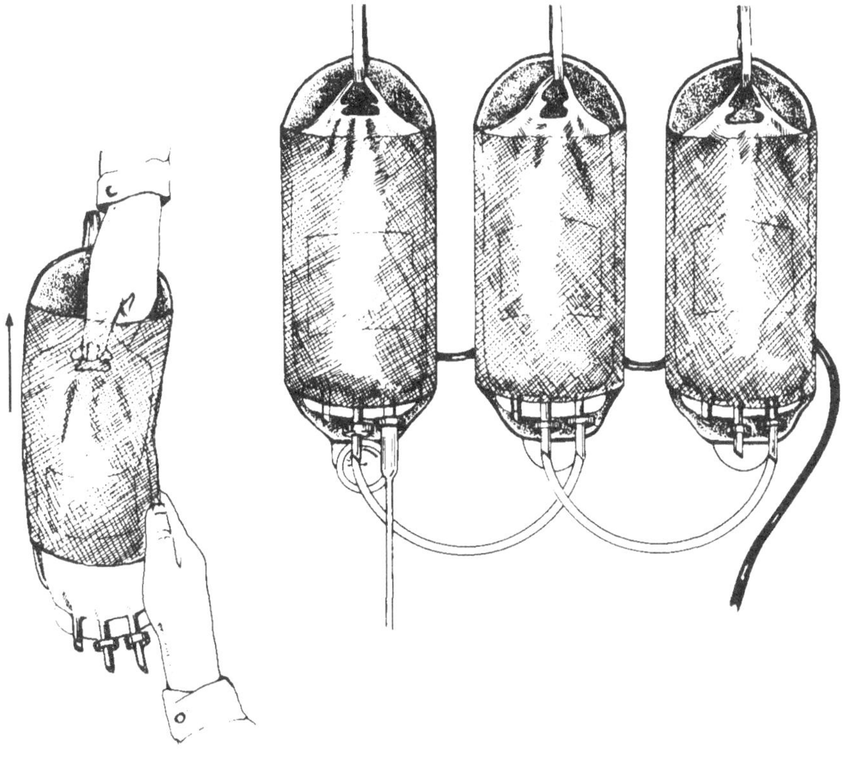

FIGURE 7. Sketch of plastic bags drawn into individual pressure infusion cuffs which are connected to a source of air pressure in order to squeeze out nutrient solutions at a low but constant pressure (about 5 lb/in.^{2})

In some institutions it is felt that mixing at home by the patient decreases the cost of the program, and allows the patient to be independent and not tied to a particular pharmacy. Training in the mixing of the solution is an intricate matter requiring great emphasis on aseptic technique. The patient must understand his own formula, the components being used, and how to use the mixing equipment. Each institution must decide which concept best fits its needs and develop an appropriate approach.

C. Pneumatic Infusion "Pump" System

This section gives details of the use of the pneumatic infusion "pump" for infusing the amino acid-dextrose solution.

The system consists of pressure infusor cuffs (Figure 7) which are pneumatically inflatable sleeves, into which the plastic amino acid-dextrose bags are placed. By inflating these cuffs to a predetermined pressure (below 300 mmHg) using compressed air, it is possible to infuse the solution at a constant pressure. The cuffs are attached to 2-stage regulators which reduce the pressure in a compressed air cylinder (2200 lb/in.^{2}) to a constant low pressure of 4 to 6 lb/in.^{2} (Figure 8).

This system has two main advantages. First, because the solutions are in bags there is no air replacement and consequently there is no danger of air embolism. Second, this is a constant-pressure device, unlike mechanical pumps which are constant-volume devices. In the event of lines being kinked at night, the system equilibrates at the predetermined pressure and there is no danger of a "blowout" of joints. With constant volume pumps the build up of pressure in a kinked line may cause a "blowout".

This whole system is attached to an intravenous pole on wheels which allows the patient to be mobile rather than being confined to bed (Figure 9).

Solassol et al., in France, utilize a portable infusion system with their ambulatory patient. The nutrient mixture is carried in a 3-*ℓ* U-shaped, silicone rubber bag. It is suspended from the patient's neck, like a halter, with a short piece of delivery tubing at its lowermost point and is connected to a three-speed miniaturized battery-driven pump. This pump weighs less than 400 g and is suspended from a waist-belt. This bag is washable and can be sterilized in an autoclave. If patients wish to infuse themselves during the night, the 3-*ℓ* bag is hung beside the bed and the solution is delivered by a rheostat-controlled pump plugged into an electrical outlet. Disadvantages of this system are its weight and the bulkiness of a 3-*ℓ* bag, the cost of manufacturing a bag of this material, and, since it is not disposable there is a risk of it not being sterilized properly. This device is not available in North America.

Dudrick and team have developed an ambulatory infusion vest, with a pocket over each breast, capable of supporting and protecting a plastic bag filled with nutrient solution. The bags are connected by specially designed Y-tubing to a miniature battery-powered volumetric pump, which is secured in a small pocket in the front of the vest.

This feeding system is said to be lightweight, but must be custom-fitted for each patient. It is relatively inexpensive and allows the patient to be ambulatory. However, the technique of intermittent overnight feeding appears to be the most commonly used approach and the one the Toronto General Hospital favors. The Hospital staff feels that a constantly worn garment and infusion system is not only a needless burden, greatly interfering with daily life and activity, but also, and quite importantly, serves as a constant and unwelcome reminder to the patient and others about him that he is dependent on a machine for survival and that something is "wrong" with him.

Scribner et al. use an infusion system mounted on a carrier, resembling a golf cart. A 2-*ℓ* bottle of nutrients and a battery-powered infusion pump are mounted on this portable cart. It has a low center of gravity and oversized wheels, which facilitate mobility.

It can be seen that as advances are made in the technology and manufacturing of equipment necessary for the delivery of HPN, procedures will benefit the patient.

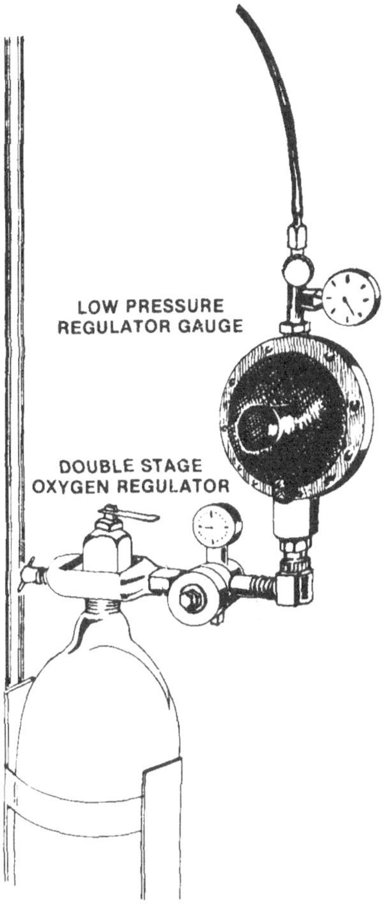

FIGURE 8. Sketch of 2-stage regulator to reduce air pressure from 2200 to 4 to 6 lb/in.[2]

1. Operation of Pneumatic Infusion "Pump" or System

1. Obtain a compressed air cylinder. Compressed air comes in a grey tank with a black and white color code on its shoulder.
2. When a full cylinder is first used the outlet should be flushed to remove any dirt in the cylinder valve. To do this:
   - Remove the protective red cap covering the cylinder outlet valve.
   - Turn the cylinder outlet away from the body (Figure 10).
   - Carefully support the cylinder, and with a tank key turn the valve ¼ turn counterclockwise. There will be a loud rush of air.
   - Turn the valve off.
3. Attach the regulator to the cylinder, lining up the correct indentations (Figure 10). If there is an "O" ring on the regulator, remove all red plastic attachments from the cylinder. If there is not an "O" ring, leave the red plastic washer in place on the cylinder.
4. Tighten the yoke screw (Figure 8).
5. Attach rubber tubing from the pressure bags to tubing on the pump.
6. Open the cylinder valve slowly with the tank key, turning it in a counterclockwise direction. The valve should be fully open.
7. Check the two gauges:

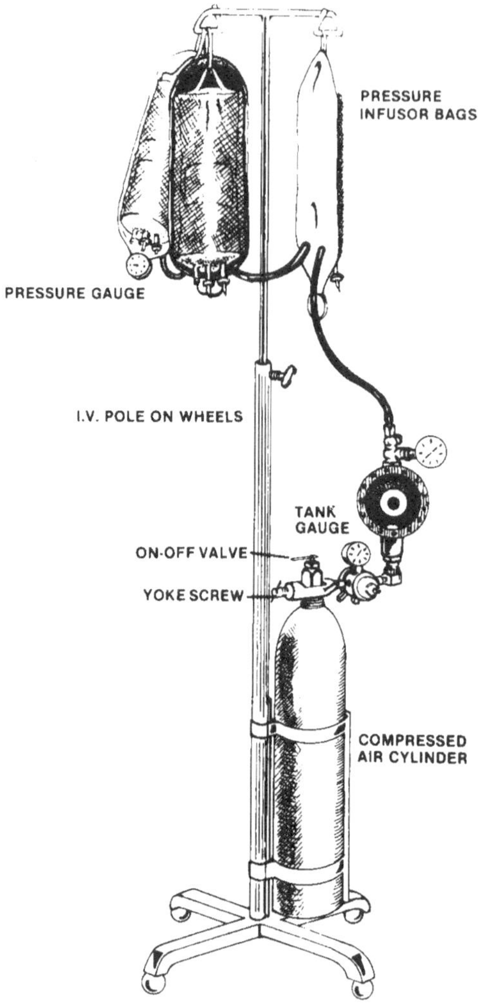

FIGURE 9. Sketch of pneumatic infusion "pump" system mounted on mobile intravenous pole; tank of compressed air, reducing valves, and pressure infusor cuffs surrounding collapsible bags of nutrient solutions. Transfer tubing from nutrient solution to catheter hub is not shown.

- When full, the tank gauge (Figure 8) should read about 2200 lb/in.^{2} This reading will decrease each time the cylinder is used.
- The low-pressure regulator gauge (Figure 8) should always be set to read between 4 to 6 lb/in.^{2}

8. The valve must remain open for the duration of the infusion.
9. Turn the cylinder off by turning the valve clockwise with the key until it stops turning.
10. Deflate the pressure cuffs by disconnecting the Luer-lock that joins the rubber tubing of the regulator to the pressure infusor cuffs.
11. To remove the regulator from the tank, loosen the yoke screw.

2. Possible Problems with the Pneumatic Infusion System

A full cylinder of compressed air (2200 lb/in.^{2}) should last for approximately 30 days when used on a daily basis. If more air than usual is being used, check:

FIGURE 10. Sketch of air-pressure tank showing outlet and indentation for alignment of the yoke attachment for mounting regulators.

1. The infusor cuffs. These occasionally weaken at the seams and a small leak may develop thus requiring more air to keep them inflated. To check for leaks, inflate the cuffs and submerge them in a sink of water. If air bubbles appear, there is a leak. Return the defective bag to the hospital and obtain a new one.
2. The regulator:
   - Check the yoke washer to make sure it is in the correct place and in good shape.
   - Check to make sure the regulator goes into the correct indentation on the cylinder.
   - Make sure the yoke screw is turned tight.

Like all machinery, the "pump" should be checked yearly to be maintained in good repair. If any problems arise with it contact the appropriate personnel in the hospital, i.e., the TPN teacher.

3. General Safety Rules for Use of the Pressure Infusion System

A full cylinder is under a great deal of pressure and can be dangerous if not handled with intelligence and respect.

1. Store cylinders by laying them flat on the ground in an area which is not prone to temperature change, i.e., away from radiators.
2. When in an upright position ensure that the cylinder is securely supported, to prevent it from falling over.
3. Check all labels and the color code on the cylinder to ensure compressed air is being used.

4. Always flush the cylinder outlet prior to attaching the regulator.
5. Never use any petroleum products in this system.
6. Never transport the cylinder with the regulator attached.
7. Keep a spare tank available in case of pressure infusor bag leaks.
8. When tank pressure gauge drops to 300 lb/in.^{2} the tank should be changed.

VI. MONITORING

Home patients should monitor their urine for glucose and ketones twice a week. The urine should be tested prior to the infusion of the amino acid-dextrose solution and again when the infusion is completed. The technique used is as follows.

A. Obtaining a Second-Voided Specimen of Urine

1. Empty urinary bladder completely.
2. Void again in half an hour to obtain specimen for the glucose and ketone test.

B. Technique for Testing Urine

The test material used is Keto-Diastix®.* This diagnostic aid gives a simultaneous determination of urinary ketones and glucose. The method of use is

1. Dip the reagent end of strip into urine specimen.
2. Tap edge of strip to remove excess urine.
3. Compare reagent side of test area with the appropriate color chart.
4. Read ketone test at 15 sec — test is positive if purple color develops. The intensity of the color is proportional to the amount of ketone bodies present.
5. Read glucose at 30 sec. Compare reagent strip with closest matching color block.

C. Misuse of Test Materials

Never use test materials that are past their expiration date, are discolored, or have been exposed to moisture. Moisture, especially, causes deterioration of the material. Always ensure that the screw-top lid is tightly on the jar between uses.

Always read the directions accompanying the test material and follow them precisely. Be sure to wait the specified length of time before reading the test result.

D. Positive Reading

If sugar or ketones are present when the urine is tested then: (1) perform urine testing technique four times a day, (2) notify physician.

VII. PROBLEMS ENCOUNTERED DURING THE ADMINISTRATION OF THE INFUSION

A. Decreased Flow Rate

Listed below are various factors likely to cause an unsatisfactory flow rate.

1. Kinks at any point in the infusion tubing, the coiled extension tube, or the Silastic® permanent catheter (especially at the hub).
2. A closed clamp.

* Available from the Ames Co., Division of Miles Laboratories, Rexdale, Ont. and Elkhart, Ind.

3. Amino acid-dextrose solution:
   - Air in the filter, the filter must be filled with solution.
   - Air leaks at any connection in the pump system.
   - Dials on the pneumatic infusion pump not within the acceptable pressure limits.
   - Incorrect position of the roller clamp — the pressure exerted upon the solution in the bags and tubing causes the clamp to roll down and obstruct flow. To mitigate this, establish the initial infusion rate at ten drops faster than that calculated.
4. Lipid emulsion:
   - Blocked air vent in tubing.
   - Vented tubing not used.
5. Clotted line (see Section IX).

B. Leak

1. Loose connection or cracked adaptor.
2. Defective tubing — if this occurs, save the tubing and return it to the hospital. The manufacturer will be informed and other home TPN patients alerted.

C. Air

1. Air bubbles may form and adhere to the plastic of the amino acid-dextrose solution bag. To eliminate them, gently shake bag and they will float to the top.
2. Air bubbles may accumulate along the walls of the tubing. To clear them, gently tap the tubing and they will rise to the top of the drip chamber. These bubbles add up to a negligible amount of air and should not cause alarm. A large amount of air in the tubing may indicate a loose or cracked connection. If so, stop the infusion and change the tubing.

D. Filter Problem

1. Occasionally, air may accumulate in the filter. To avoid this, flush the line completely and correctly the first time. However, if air does accumulate during the infusion and the drip rate decreases, change the filter.
2. The main purpose of the IVEX-2® filter is to prevent air passing beyond it. If air does go through it, change the filter and return the defective one to the hospital.

E. Disconnected Tubing

The tubing may become disconnected if it is not assembled correctly or taped securely. If disconnection occurs the patient should:

1. Remain calm — DO NOT PANIC.
2. Maintain sterile technique.
3. Clamp the permanent Silastic® catheter with padded forceps to prevent air entry.
4. Mask, and wash hands.
5. Remove contaminated tubing from permanent line.
6. Follow the declotting procedure (Section IX, A.2).
7. Close the clamps on the transfer sets. Clamp porthole 1 with forceps so the fluid will not leak out. Remove contaminated tubing and hang new tubing and filter with extension tubing.

8. Retape connections and resume infusion.
9. If there is only a small amount of solution left, i.e., less than 100 mℓ/bag, discard the whole setup and heparinize the permanent catheter (see Section VIII, following.)

VIII. HEPARIN LOCK

In order to render the patient free of the infusion system for a significant part of a 24-hr day, a heparin lock is utilized.

As previously stated, the patient is infused over an 8 to 10 hr period during the night, and is then detached from the infusion apparatus. After disconnection, the line is kept patent until the next "hookup" by instilling heparin. The patient can now move about and pursue daily activities.

A. Heparin Lock Techniques

1. Assemble

- Mask
- Heparin (1000 units/mℓ)
- Alcohol
- Medication cap*
- Adhesive tape
- 25-Gauge needle

2. Procedure

1. Mask and wash hands.
2. If a 75-cm (30″) primary extension tube is used, draw up 7000 units (7 mℓ) of heparin OR if a 17-cm (7″) primary extension tube is used, draw up 4000 units (4 mℓ) of heparin.
3. Clamp intravenous tubing and extension tubing.
4. Clean the connection between the intravenous tubing and extension tubing with an alcohol swab and then disconnect the two.
5. Pinch the extension tube to raise the fluid level. Attach a medication injection cap (Figure 11).
6. Swab the cap thoroughly with alcohol.
7. Open clamp on extension tube.
8. Instill heparin slowly through the cap.
9. While withdrawing the syringe begin to close the clamp on the extension tube.
10. Check that the clamp is completely closed and the medication cap is securely attached, and tape.

B. Heparinizing Procedure When Infusion is Given Discontinuously

If the amino acid-dextrose solutions and lipid emulsion are administered discontinuously, i.e., lipid emulsion from 12 noon to 2 p.m. and amino acid-dextrose solution at 10 p.m., the line must be heparinized between the infusions. The technique is modified as follows:

1. If the line is to be heparinized after the lipid emulsion has been infused, first flush the line with 10 mℓ normal saline.
2. Inject heparin according to the heparin lock technique (Section VIII, A).

* Available from Becton Dickinson, & Co., Mississauga, Ont.

FIGURE 11. Sketch of medication cap with Luer-lock used to plug end of catheter when patient is disconnected from infusion system. Cap allows insertion of syringe needle.

C. Heparinizing Procedure When TPN is Administered Other Than On a Daily Basis

If TPN is required only once or twice per week, the heparinization routine is

1. In the morning, instill 7000 units (7 mℓ) of heparin into the extension tube following the heparin lock technique (Section VIII, A).
2. In the evening, 12 hr later, instill 2000 units (2 mℓ) of heparin, again following the heparin lock technique (Section VIII, A).
3. This procedure is followed until the night of the infusion, then 2000 units (2 mℓ) of heparin is omitted.
4. In the morning after the infusion, this procedure is repeated.

D. Heparinizing Procedure — For Use of Additional Medication

Occasionally additional medication, e.g., antibiotics, is required. These are administered through the permanent line, several times per day, and must be followed with a diluted heparin solution. The procedure is

1. Mix 3000 units (3 mℓ) heparin with 7 mℓ of normal saline in a 10-mℓ syringe.
2. Draw up 10 mℓ of normal saline in a new 10-mℓ syringe and administer through the line.
3. After the medication has been given, flush the primary extension tube with 10 mℓ of normal saline.
4. Instill 7 mℓ of the diluted heparin solution.
5. Every 4 hr instill 3 mℓ of the heparin dilution until the medication is due.
6. When the medication is to be administered again, the line must be flushed with 10 mℓ of normal saline, the medication given, followed by 10 mℓ of saline and then the 7 mℓ of the heparin dilution.

IX. CLOTTED LINE

With appropriate care the line should not clot, however if it does, the following declotting procedure should be carried out.

A. Declotting Technique

1. Assemble

- Sterile bottle normal saline
- Heparin (1000 units/mℓ)
- 2 10-mℓ Syringes
- 1 Tuberculin syringe (1 mℓ)
- Extension tube
- Medication injection cap

- Alcohol swabs
- Padded clamps
- Mask

2. Procedure

1. Mask and wash hands thoroughly.
2. Draw up 7 mℓ of normal saline in the 10-mℓ syringe.
3. Clamp permanent line with padded forceps.
4. Attach the syringe (no needle) to the hub of the permanent line. Unclamp line. Attempt to aspirate (pull back) the clot.
5. If unsuccessful, draw up 1 mℓ of heparin (1000 units/mℓ) in the tuberculin syringe. Clamp line. Attach the syringe to the hub and then unclamp the line. Attempt to aspirate the clot again.
6. If nothing returns in the syringe, instill the heparin into the line. Leave the syringe attached to the hub and clamp the permanent line for 10 min.
7. After 10 min attempt to aspirate the clot.
8. If this is successful, flush the line with 500 mℓ of normal saline and then heparinize the line according to the heparin lock technique (Section VII, A).
9. Notify the physician.
10. If the above is not successful, place a new medication injection cap on the hub and come to the hospital immediately.

3. Declotting With Urokinase

1. Instill into the catheter 7500 IU of urokinase* diluted in a total volume of 3 mℓ of sterile saline.
2. Clamp the catheter with padded forceps for 3 hr, leaving the empty syringe on the end of the catheter.
3. Flush the catheter with 1000 units heparin diluted in 19 mℓ of saline.

X. PRIMARY EXTENSION-TUBE CHANGE

The primary extension-tube should be changed every seven days. There are two ways this may be done.

A. Method 1

1. In the evening of the day the tube is to be changed, attach a new primary extension-tube to the tubing system used to infuse the amino acid-dextrose solution (see Section V, B).
2. Flush air out of the tubing.
3. Clamp the permanent catheter with padded forceps.
4. Swab the connection between the blue permanent hub and the old primary extension tube with alcohol.
5. Disconnect, pinch the ends of both tubes to raise the fluid level, and connect the new primary extension-tube to the blue hub.
6. Remove forceps and begin administration of solutions.

* For additional details see Glynn, Langer, and Jeejeebhoy, 1980.

B. Method 2

1. Remove a new primary extension-tube from package.
2. Carefully draw up 10 mℓ of normal saline in a 10-mℓ sterile syringe.
3. Attach the syringe, without the needle, to the end of the primary extension-tube that does not connect to the permanent blue hub.
4. Flush air out of the tube.
5. Clamp permanent catheter with padded forceps.
6. Swab connection between the blue permanent hub and old primary extension-tube.
7. Disconnect, pinch ends of both tubes to raise the fluid level, and connect the new primary extension-tube to the blue luer-lock hub.
8. Close slide clamp. Remove syringe on end of extension tube and attach a blue medication cap.
9. Remove forceps on permanent line.
10. Follow heparin lock procedure (Section VIII).

XI. POSSIBLE PROBLEMS ASSOCIATED WITH HOME TPN

A. Infection

Absolute and meticulous care must be taken to prevent any type of infection. Since the permanent catheter remains in place on a long-term basis, the entry site must be cleaned and dressed well to prevent the growth of any organisms. Contamination of the TPN solutions or a break in the sterility of the infusion apparatus or its connections must be avoided. These could promote growth of organisms down the catheter to the tip, necessitating removal of the catheter.

If symptoms occur, i.e., fever (over 38.5°C), chills, lethargy, and weakness, notify the physician and he will advise treatment. For example, urokinase followed by a 4- to 6-week course of antibiotics through the permanent line may be required. Conventionally a line infection was treated by removing the catheter. However, recent experience by Glynn and associates has shown that most infections, especially those due to *Staphylococcus epidermitis,* can be eradicated by instilling urokinase followed by a 4- to 6-week course of antibiotics given through the line. With *S. epidermitis* a cephalosporin such as Keflin® or Cloxicillin®, given in doses of 2 g every 6 hr will eradicate the infection in about 70% of patients. This approach has not been associated with any complications. The Keflin® or Cloxicillin® is given as a rapid infusion in 100 mℓ of 5% dextrose in water through the central venous line just after stopping the infusion in the morning and again in the afternoon, followed by a continuous infusion added to the TPN solution through the night. The dose added should then be double that given on each occasion during the day. As an example, 2 g of Keflin® will be infused at 8 a.m. and 2 p.m., and 4 g added to the TPN solutions for overnight infusion.

B. Back-Up of Blood in the Extension Tube

A small amount of blood may appear in the primary extension tube if there has been an increase of pressure in the subclavian vein. This pressure increase may be caused by any form of exercise, bending over, etc. This phenomenon is not a cause for alarm, for the heparin in the catheter and extension tube will prevent any clotting. If excessive blood appears, flush the line with 2000 units of heparin (2 mℓ) (see Section VIII) and close the clamp completely on the extension tube.

C. Hypoglycemia

Hypoglycemia occurs when there is a sudden decrease or a deficiency of sugar in the blood. If the amino acid-dextrose solutions are infused quickly and then suddenly discontinued, symptoms of headache, nausea, nervousness, cold sweats, blurred vision, and drowsiness may appear. If this is not treated, seizures or coma may result. This may occur while the patient is "hooked up" or shortly after discontinuing the amino acid-dextrose infusion.

If oral intake is permitted, juice with additional sugar must be consumed.

If oral intake is not appropriate, i.e., poor absorption, infuse a bag of amino acid-dextrose solution until the symptoms subside. The infusion should be discontinued gradually by decreasing the drip rate. The remaining solution is discarded.

D. Hyperglycemia

This condition may occur if the infusion of the amino acid-dextrose solution is administered faster than the body has the ability to utilize this glucose. Symptoms include an increase in frequency and volume of urine, thirst, presence of sugar in urine (see urine testing procedure — Section I in Chapter 7). If these symptoms occur:

1. Gradually decrease the flow rate of the infusion and void to empty the bladder.
2. Recheck the urine sugar in 2 hr.
3. If the sugar is still present in the urine despite slowing the rate of the infusion, or if the rate at which the infusion has to be infused is slower than desirable, call the physician.

E. Air Embolism

This occurs if air is allowed to enter into the venous system of the body's circulatory system. All tubings must be completely flushed of air prior to use. Whenever the permanent line is opened, i.e., changing extension tube, etc., it must be clamped with padded forceps to prevent air entry. The design of the home TPN system excludes the possibility of air being pumped into the bloodstream. However, if air does enter the catheter watch for chest pain, shortness of breath, and coughing. If present the patient should:

1. Clamp the permanent catheter securely.
2. Lie on his/her left side.
3. Inform the physician.

F. Fluid Overload

If large volumes of solution are infused or are administered rapidly, the circulatory system may be affected. Signs of weight gain and peripheral edema (accumulation of fluid and shortness of breath) indicate a problem and the physician should be notified.

Patients that have been malnourished for any length of time may have low serum protein levels and thus will experience peripheral edema. As their nutrition improves, this "refeeding edema" will subside.

G. Dehydration

This process occurs when the output of body fluid exceeds fluid intake. This may be due to the deprivation of water, excessive loss of fluid through G-tubes or ostomies, reduction in the total quantity of electrolytes, or because of the infusion of hypertonic solutions.

These situations can be precipitated by hot weather, gastroenteritis, flu, ingestion of alcohol, etc. If signs of thirst, dry mucous membranes, poor skin turgor, weakness, or dizziness occur, notify the physician.

H. Oral Hygiene

Good oral hygiene must be maintained at all times. Teeth should be brushed 3 or 4 times a day and mouth wash and dental floss used regularly. The gums should be exercised and massaged. This will prevent oral infections such as cankers or parotitis caused by the collection of bacteria. Regular dental checkups will help. Patients with limited oral intake should especially adhere to this. If dental work has to be done then prophylactic antibiotics should be given. The recommended regimen is Cloxicillin® just prior to the procedure and 2 hr later. If patient is allergic to penicillin then Vancomycin® is given.

I. Malaise or Illness

If the patient does not feel well he should never hesitate to contact the physician. A fever or slight temperature elevation can be of concern and the physician should be notified.

XII. TEACHING THE "BACK-UP" PERSON

In addition to teaching the patient, it is desirable to train another person in the ways of managing the home TPN system. This person may be a spouse, family member, or close friend. In any case, such a person helps the patient and provides assistance if problems arise preventing the patient from carrying out any of the necessary technical maneuvers.

When the patient's own training has been completed, the designated assistant or "backup" person should be taught by the patient. This method is valuable for two reasons. First, it allows the nursing staff to evaluate the patient's level of knowledge, and thus the effectiveness of their teaching. Secondly, it increases the patient's confidence in the understanding of the system.

It is stressed that the patient is the one responsible for the system and the other person is available to offer encouragement, reinforcement, and to help when required — not to act as a constant crutch.

XIII. THE EFFECTS OF HOME TPN ON LIFESTYLE

The effect of home total parenteral nutrition on a patient is best illustrated by the experiences of one such patient who has lived with this system for more than ten years.

On September 23, 1970, after 3 days of pain, I left home and 3 children for a *quick trip* to the doctor's office. That *quick trip* ended on July 11, 1971. After an examination we were sent over to Scarborough General Hospital and the operating room. Four inches of bowel were resected and that should have been the end of things, but in about ten days I was in the operating room again having more surgery which left me, thanks to gangrene, with a duodenal stump and a rectal stump, and the doctors with a rather large problem — how to keep me alive. Fortunately one of the doctors had worked with Dr. Jeejeebhoy who was good enough to take on the challenge at the Toronto General Hospital.

A great deal of the first month or so at T.G.H. is a jumble of hallucinations and reality. One of the bits of reality comes to mind when I'm asked *What was it like when they told you you couldn't eat any more?* Lying on the operating table, under a local anesthetic, I heard one of the surgeons say *If we don't find a vein for this line, she'll die.* Eating somehow didn't seem too important. Fortunately a vein was found. With the co-operation of a number of specialists a permanent site was found and we were on our way.

The next months were days and nights of trial and error and playing guinea pig for tests of every conceivable kind. If it hadn't been for that tremendous support and loving care of my family and the staff I don't think I'd have made it. Surgery after surgery was the order of the day. Months in bed connected to tubes, pumps, bottles and of course, my feeding were part and parcel of those days. Even the *Vampires* who came every morning — and sometimes during the day — even at night — to drain every bit of blood I ever had, became good friends.

Dr. Jeejeebhoy was far-sighted enough to know that my greatest worry was over Cliff, my husband, and our children, so the rules were bent and the kids were allowed to visit every Sunday afternoon. What an experience for them! They had had almost no hospital contact, so that to come to a strange room in a strange building to see their mother, always healthy and on the go, lying in bed, attached to all these tubes, bottles, etc. was positively terrifying. It was a few weeks before Miriam, age 7, could stay for more than five minutes before having to be rushed out to the fresh air to keep her from fainting. Children adapt to situations much more quickly than adults however and it wasn't so very long before the Sunday visits were part of the weekly routine and all the happenings of the week were stored up for telling on that day. Even Christmas, gifts and all, was celebrated in that little room at T.G.H. and Christmas dinner with all the relatives, was *eaten* in the cafeteria.

Just before New Year, I was allowed out of bed. What a production! Disconnecting the various tubes from their pumps and attaching them to bottles which were then put into green garbage bags and with a nurse on either side to support me, two more carrying the garbage bags and one in front to push the intravenous pole, we had our first little stroll into the hall. Out into the hall and back and I was exhausted, but after that it was nearly an every day occurrence. The time finally came when there were fewer bottles and then none at all — just clamps to turn off when I was up, and I could get around by myself.

During this time however there were many setbacks. There was the time, when I lay in bed for a certain length of time every day feeling *awful*. Legs and arms like lead, head aching, and for no apparent reason. After several days the tribe (as I called them — doctors, residents, interns, nurses, etc.) came into the room and stood around the bed grinning — apparently there was too much alcohol in my solutions* and I was suffering from a hang-over every day! This of course was rectified immediately but we still laugh about it. Another problem was abscesses. Things would just begin to move ahead, when suddenly I'd spike a fever and we'd know another abscess had formed and it would be back to bed. Stitches were another problem — they would let go in spots and until you have had the sensation of actually looking down into your own insides, you cannot begin to understand the feeling. Finally they sewed me up with wire — which seemed to do the trick. Blocking of the Silastic catheter was another problem. The nurses spent many a nerve-wrecking half-hour moving a clot, or drawing out air from the line. That was a scary time for all of us.

As time went on I spent many hours up and down the hall visiting the other patients, especially at meal time. It is very important for a TPN patient to be around food and the smell of food because it is something you have to deal with every day of your life. Luckily not all are in my position of not being able to eat at all. For someone who liked to eat, the way I did, it is a temptation which really needs work if it is to be overcome. Eventually the day came when they asked if I'd like a tray — all liquid — of course. How thrilled I was! The first five months I was only allowed ice chips — then ice water — now a tray! The following morning my tray was brought in and the first thing I saw was lime Jello. I suddenly lost my appetite. Lime is one of my favourite flavours but never in Jello. They didn't need to worry about me overeating.

By this time I had been taught to change my bags of feeding but although I thought I was doing very well at it, a group of students and their teacher beat a hasty retreat when I was going to do it in their presence. I had also been taught to do my dressings. The lessons were very thorough with the idea in mind that I could do this when I went home (if that were possible). Disconnecting me from my feeding so that I would have a certain amount of time each day without my intravenous pole had finally been accomplished after a lot of thought and a couple of unsuccessful attempts, but no system had yet been found acceptable so that I might go home.

Finally at the end of June, Dr. John Wright and a young man from the Inhalation Therapy Department called me into the utility room to introduce me to *Lester*.** This *creature* consisted of an intravenous pole with a compressed air tank attached and three pressure cuffs attached in sequence. Into these cuffs the bags of feeding were to be inserted, the cuffs attached to the air cylinder and then air turned on to a preset pressure. The bags of feeding, also attached in sequence, would then, with a regulated drip — be taken at night while I slept.

With the advent of Lester everything fell into place and in 10 days I was home. What a home-coming that was. When I had become ill we were living in Scarborough — now, with Dr. Jeejeebhoy's permission I was going home to what had been our cottage, 100 miles away, near Bobcaygeon. Cliff had altered what had been just a cottage shell into a beautiful home. I had been consulted while in hospital, on rugs and panelling. The kids had *Welcome Home* signs and flowers everywhere, and yet it was only recently that they had been terrified of having me come home and had even discussed the problem with their father.

That first night will always remain in my mind, and in Cliff's as well. Granted, we had both been well trained in all procedures, but suddenly we were 100 miles miles away from help and we were both extremely nervous but putting up a good front. Neither of us slept much the first few nights, but gradually we relaxed and accepted the fact that *Lester* performed just as well at home as he did in hospital. The first months at home meant being *hooked up* approximately 16 hours a day at first, but eventually it was cut down to twelve — 9:00 p.m. to 9:00 a.m. Now it is about 10 hours.

* In 1969 when this happened the Toronto General Hospital was using a commercial TPN mixture, Amigen® 800, containing ethanol!

** In those days of 1970 high volume pumps were not available and the system used has stood the test of time.

We didn't change our lifestyle very much as we were never gadabouts. We still had people over for cards and went out if we were asked. I simply changed my Intralipid to a morning period instead of a night period. Friends and neighbours were most embarrassed at first. In the country, a lunch is served at every *do* and it took them quite a while to relax and eat when I was around. Now they forget to the point of offering a plate to me as they go by and then apologize for forgetting.

The family adjusted very well and very quickly and it wasn't long before they were bringing their friends home to watch the setting up of my feeding. Miriam, in grade 4, did a science project on the procedure — even taking the empty bags and setting them up on bristol board. Her teacher had to take it to his sister, my G.P., to confirm that everything was right.

Some problems did arise. A clot in my line, sent us first to Dr. Junkin and then to Toronto. The movement of the line required a trip to the hospital and the insertion of a new line — which I still have. I became *diabetic* for a short time but upped my chromium intake and all symptoms disappeared. My G-tube has had to be reinserted twice other than the regular changes. Both times we have gone to the nearer Peterborough Civic Hospital, where it has been done with no trouble. The odd fever and a bout of pneumonia have required hospitalization but on the whole I have spent very little time in hospital. I accompany Cliff when he picks up my supplies once a month. While he does that I try to see any new patients for TPN. There are so many questions and — that was something I missed — no one could give me any answers. I try to stress the fact that there is no need for TPN to run your life as long as you obey the rules. Since we came up north I have learned to drive, I taught a 4H-group for six years, sing in the church choir, bowl once a week and work one morning a week.

There are others in my position, some of whom hold full-time jobs. We have formed a group of all TPN patients which meets every six months to keep up our friendship, answer each other's questions and be of any help we can be to one another.

After 10 years I feel very fortunate to have TPN and Dr. Jeejeebhoy, and the ability to lead a very full and *normal* life.

Now, taking each aspect of lifestyle, we can make the following generalizations.

A. Meals

While patients should not be prohibited from eating for psychological reasons, eating may have to be restricted or may not be possible due to bowel disease. Every patient not allowed to eat or who cannot eat misses and craves food. They miss the demarcations of the day which mealtime has always provided. This monotony of the TPN routine deprives them of a diversion that human beings universally depend upon. Also, they can no longer turn to the refrigerator to relieve boredom, anger, and frustration. Eating is a normal function of everyday life, a major aspect of the socialization process and an essential family event.

In early stages, the absence of food may cause depression, but the degree of depression diminishes as patients and their families adapt.

Most female TPN patients continue their role as the "cook" and some men even assume this responsibility if temporarily unemployed. Many patients sit with their families at mealtime and may even begin their infusion.

B. Bathing

Routine bathing is permissible. A shower or bath may be taken. The bather should prevent the occlusive dressing over the permanent catheter from becoming excessively wet. For example, when in the shower the bather should stand with his/her back to the spray, preventing a stream of water from hitting the dressing and permanent line.

C. Sleep

Since the intravenous feeding occurs during the night, this may lead to a disturbance of the sleep pattern. Due to the high fluid input over 8 to 10 hr most patients must rise 2 or 3 times a night to urinate. In addition, initially some are concerned about proper functioning of the equipment, i.e., rate of solution infusion, placement of permanent catheter, etc.

Upon rising the patient should check the drip rate of the solution. Most patients have little problem returning to sleep although this disturbance may cause fatigue and necessitate a nap during the day. If excessive frequency is experienced this should be reported to the physician, who should evaluate the fluid load and consider reducing it.

D. Exercise

Patients should be encouraged to return to their normal exercise activities. Exercise will promote protein synthesis forming muscle rather than fat deposits. All sports are allowed except for swimming and "contact sports" such as football. An exercise regimen should be gradually progressive in vigor. Some will find they can do more since their last illness, while others will feel their limitations are greater. Catheter placement should be adapted to the eventual sport enjoyed by the patient. For example, in hunters the catheter should be placed on the left side or inserted medially through the jugular. This keeps the catheter away from the site of pressure on the shoulder due to the recoil of a firearm.

E. Employment

One of the objectives of the home TPN program is to allow a person to resume a normal lifestyle. It is only natural then to assume that one would return to a work situation. Patients are encouraged to return to work as soon as their physical and emotional condition permits. If they work a regular 8-hr day they must learn to schedule their feedings around this. Self-employed or part-time workers may have a more flexible schedule. Shift work is not advised.

F. Body Image and Sexual Activity

Most home TPN patients have had extensive surgery and may have a severely altered or negative body image. Some have difficulty in accepting their multiple scars, as well as the actual placement of the permanent catheter. They may perceive themselves as unattractive or repulsive, which will affect their self-respect and self-esteem.

This, in turn, may or will affect the patient's attitudes and sexual life. Their greatest fears and anxieties are related first to dislodging the catheter with excessive motion or varied position and second, to appropriate timing. They feel that lovemaking must now be "scheduled", lacking spontaneity.

Patients should be reassured and encouraged to resume sexual activity. They should experiment and try to adapt their sexual lives to include their line. Many choose to have intercourse when they are not "hooked up".

Single and younger patients, who have spent a large part of their life time in hospital or at home, have concerns about being isolated and socially unacceptable. A prime need here is to maintain or improve their self-esteem.

G. Entertainment

Gradual contact with the community should be encouraged. Utilization of TV, radio, newspaper, and telephone provide this first link of communication. When feeling well enough, contact with relatives, friends, colleagues, etc. should be considered. Solution administration schedules can be adjusted easily. For example, if an evening out is desired when the infusion of the lipid emulsion is usually done, then it may be omitted and given the next morning following the infusion of the amino acid-dextrose solution. Also, the amino acid-dextrose solution may be prepared and mixed, the tubing inserted but not flushed, and all refrigerated prior to going out.

H. Travel

Various types of vacations or trips may be planned when the patient feels comfortable and secure with all the procedures and routines of the home TPN system.

A great deal of organization is required to make this possible, but it certainly is feasible. When a vacation is planned, notify the physician and request a letter approving this travel. Notify the agency, i.e., airlines, train, etc., of the situation. Pack supplies carefully. Contact another home TPN patient who has had experience in this.

XIV. EMOTIONAL AND PSYCHOLOGICAL ADJUSTMENT TO HOME TPN

The early stages of adjustment to home TPN are fairly demanding upon the patient. Once the patient has attained a satisfactory degree of physical health and is familiar with the necessary technical skills of the TPN system, it then becomes apparent that there must be certain psychosocial adaptations.

All patients are individuals and regard their hospitalization as unique experiences. Each one will have a diverse range of feelings and emotion about the home program and will require encouragement to verbalize, discuss, and resolve these feelings. This is an important part of the role of the nurse as well as other members of the team. If this does not occur, unresolved anger, denial, and depression may affect the patient's total acceptance of the program.

The sequence of stages through which patients may progress is similar to those in Kubler-Ross's stages of death and dying especially in patients who were previously healthy and in whom a catastrophic illness led to the need for home TPN. Initially, there is a reaction of disbelief and denial in which the patient cannot believe this is happening to him, with the likelihood of permanence of this mode of therapy.

This is followed by feelings of depression, sorrow, anger, and grief, with preoccupation about a probably greatly altered way of life. These include the need to master a complicated, artificial system and the need to change eating habits. When the need to restrict eating is very severe then the psychological problems are greater, for eating is a function of our daily and social lives. Many of our human encounters involve food, i.e., coffee breaks, eating in restaurants, holidays, etc. and there may be only slow acceptance of the fact that eating is now compromised.

When this has been worked through, there is a gradual process of reorganization in which the patient tries to incorporate his loss, accept his limitations, and adapt to a new way of living. This final stage is likely to be accelerated by being able to go home sooner.

In contrast, patients who have suffered from a severe chronic illness, incapacitating them and necessitating many hospital admissions in the past (as with Crohn's disease), adjust more easily and often find a new life with home TPN. Such patients find that what they must sacrifice is worth the experience of improved health and more energy to pursue activities of daily living.

In patients with both acute and chronic illness requiring home TPN there is a high degree of anxiety and fear of the unknown. They may be overwhelmed with the technique and knowledge required to care for "their lifeline" and the uncertainty of what will occur in the future.

Various resources should be available to assist the patient in coping with these feelings. An effective method is the introduction to a "veteran" home TPN patient. This allows for an exchange of ideas, concerns, suggestions, and methods of coping. Nursing and medical personnel should be available to listen and talk, to clarify and dispel misconceptions, to encourage questions and verbalization and to provide information and support.

The home TPN program has ramifications for members of the family unit as well. The whole system may overwhelm them or perhaps they are fearful of how their future may be changed. Time must be spent with the members of the family to help resolve these fears, thus allowing them to assist the patient freely in his adaptation.

XV. DISCHARGE PLANNING

Plans for discharge must be a joint effort between the patient and all members of the TPN team. Planning for going home begins when the training program commences and continues until the patient is home.

A. The Team

1. Doctor's Role

The doctor should evaluate the patient's physical and psychological readiness for going home. Together with the patient and other team members, a discharge date can be determined.

2. Nursing Role

Nursing staff should encourage the patient to express feelings about going home. They should be available to answer questions and help solve problems. There will be concerns about finances, returning to work, sexual experiences, and feelings of guilt, anger, fear, etc. It is important to establish and maintain a good communication pattern with the patient and family.

The physical layout of the patient's home, storage of supplies, where to change the dressing and mix the solutions — all should be discussed with each patient. For example, the dressing should be changed in a quiet, clean area, in close proximity to the stove and sink to decrease the risk of infection and/or error. The nurse plays a large part in the decision of discharge by deciding whether the patient is psychologically ready and all skills meet the standards.

3. Pharmacy Role

Each home patient is assigned to a pharmacy technician who will assist the patient in ordering supplies (see Chapter 12) compile this order and then greet the patient when the supplies are picked up.

Prior to discharge, this technician visits the patient on the nursing unit and discusses the order list. The patient is then taken on a tour of the Pharmacy department to see how the amino acid-dextrose solutions are made, where the supplies are kept, and how to go about picking them up, i.e., where to park, etc.

4. Social Worker Role

This team member can assist the patient in solving various types of problems. Education of the patients and their families about the availability of community resources appropriate for their needs is necessary. Sometimes budget and debt counseling is required due to the financial strain of additional medical bills and reduced income. Vocational and educational rehabilitation may be necessary. The social worker can be of substantial help to the patient in such areas.

5. Patient's Role

Prior to going home the patient is requested to purchase several items. These are

1. A large roasting pan to sterilize the stainless steel dressing tray.
2. Two kitchen tongs used to remove the sterile tray from the pan.
3. A refrigerator to store the TPN solutions and additives. A separate refrigerator is needed so that a supply for a month can be refrigerated and so these supplies will not be contaminated by the family's food. The refrigerator should be 13 to 15 ft^3 in size and have a constant temperature of 4 to 8°C.

The home itself does not require much alteration. Space is required for the additional refrigerator, but the location is optional. Cupboard or shelf space is needed to stock supplies of needles, syringes, tubing, cleaning solutions, etc. A neat and orderly arrangement will make monthly inventory and ordering easier.

The patient must feel comfortable and fully prepared to be discharged home. Discussions with the team will aid this.

B. Discharge Criteria

By the end of the training program, the patient should meet the following criteria:

1. Exhibit a basic understanding of the theory of TPN related to their problem.
2. Demonstrate aseptic and technical skills required to employ the system at home.
3. Develop a problem-solving approach.
4. Display an initial emotional and psychological adjustment to this new way of life.

The appropriate person (head nurse, TPN teacher) will review the total program with the patient and back-up person before discharge. All three must be satisfied with the outcome of the session before the patient is allowed to go home.

XVI. FOLLOW-UP AFTER DISCHARGE

Continuity of care necessitates that the hospital personnel continue their contact with the discharged patient. This bond gradually can be weakened when the patient feels comfortable and secure. Initially, the individual may encounter unpredictable and unexpected problems and require some assistance or advice. When this need arises, the patient can contact any member of the team.

In order to enhance the transition from hospital to home, use of a community visiting nurse service is employed. She may visit daily at first and then taper off these visits as circumstances dictate. This is a decision made between the nurse and patient. This nurse is present in a supervisory capacity. The nurse should observe the patient's aseptic technique, be aware of complications, and make suggestions that will allow an easier adjustment.

A nurse from the hospital, i.e., TPN teacher, should attempt to visit the patient 1 week after discharge. At this time, problems can be discussed and technique in the home setting can be observed. Other community resources such as the family doctor or community nurses may wish to be present at this time. This is an excellent opportunity for an exchange of information. The patient returns for medical follow-up approximately 3 weeks after discharge and then at 6-month intervals, unless there are problems.

A. Helpful Hints

The following is a list of suggestions that will help to make the home TPN system more manageable.

1. Equipment

1. Apply carpet castors to the intravenous pole. This will allow easier movement of the pole from room to room, especially if there is carpeting.
2. Place self-adhering hooks to the walls throughout the house in inconspicuous places. Rather than pushing the intravenous pole around, the bottle of lipid emulsion can be held above the head while carrying it in order to hang it on one of these hooks.

3. Purchase a second-hand refrigerator for the TPN supplies that must be refrigerated, rather than a brand new one.
4. Arrange supplies in a cupboard or on a shelf in an organized fashion. Label the shelf with the product and the amount of each the patient requires for a month. This will make the ordering of supplies much easier.
5. Keep a flashlight at the bedside. Using this during the night to check the drip rate of the amino acid-dextrose solution or if one must get up to use the bathroom.

2. Emergencies

1. Compile a kit containing a mask, alcohol swabs, normal saline vial, heparin, syringes, padded clamp, medication cap, and tape. This should be carried with the patient wherever he/she goes.
2. Keep a similar type kit at the bedside.
3. Obtain a note from the physician explaining the medical condition and how it is being treated. This will enable one to obtain medical supplies, if needed, when away from home.
4. Obtain a Medic Alert bracelet stating the medical condition and the presence of a permanent catheter.
5. Refer to the manual, checklist, and instruction sheet to help solve problems.

3. Travel

1. Construct a collapsible intravenous pole to use when traveling.*
2. If only going on short trips, store solutions in a cooler or in a portable refrigerator that plugs into the cigarette lighter receptacle in the vehicle.
3. Remove tubing from boxes and store in clean plastic bags.
4. Consider using a disposable dressing tray to change dressings while away.

4. Medical Treatment

1. If any type of surgery or dental work is required, notify the physician. A course of antibiotics will be required as prophylaxis.
2. If hospital admission is required one should bring the pneumatic infusion "pump" system, pressure infusor bags, and a 5-day supply of the amino acid-dextrose solution.

READING LIST

Dudrick, S. J. et al., New concepts of ambulatory home hyperalimentation, *J. Parenteral Enteral Nutr.*, 3, 73, 1979.

Fischer, J., *Total Parenteral Nutrition*, Little, Brown, Boston, 1976.

Scribner, H. and Cole, B., Evolution of the technique of H. P. N., *J. Parenteral Enteral Nutr.*, 3, 60, 1979.

* The authors may be contacted for details.

Chapter 12

THE ROLE AND RESPONSIBILITIES OF THE PHARMACIST IN TOTAL PARENTERAL NUTRITION

George Tsallas

TABLE OF CONTENTS

I. INTRODUCTION

The rapidly developing area of total parenteral nutrition (TPN) creates many opportunities for the hospital pharmacist to participate as a member of the "TPN team". The hospital pharmacist is an expert in aseptic manufacturing techniques and knowledgeable about various TPN solutions, making him or her an important member of the nutritional support team.

The responsibilities or functions of the pharmacist will vary with the hospital, depending on the scope of the program and the extent of the pharmacist's involvement in this program. The various areas in which the pharmacist can assume a responsible role are outlined herein.

II. SCREENING AND INTERPRETING ALL TPN ORDERS

The pharmacist must assume responsibility for carefully screening all orders and interpreting for potential problems, such as incompatibilities, inaccurate calculations, incomplete directions, and inappropriate doses.

III. FORMULATION AND PREPARATION OF SOLUTIONS

The pharmacist has the knowledge and specialized training in aseptic technique, physiochemical incompatibilities, good manufacturing procedures, proper use of equipment, and the ability to calculate dosages, thus making him/her a suitable person to be responsible for the preparation and quality control of the TPN solutions. He/she may assist the physician in making formulations that will meet general and special requirements. In addition, the pharmacist may be of help in giving advice about compatibilities, composition, and compounding of solutions.

IV. RELEASE OF THE FINISHED PRODUCT

The pharmacist bears the ultimate responsibility for inspecting the finished solution. He/she must check and confirm that the final product has been compounded in accordance with the Master Formula and all necessary records completed. The pharmacist must double-check all containers of all ingredients used, and visually inspect the physical appearance of the admixture. He/she must see that the label is complete and accurate. Furthermore, the pharmacist must ensure that the necessary quality control measures have been undertaken prior to the release of the product.

V. TEACHING AND TRAINING

The pharmacist has a role in the teaching of both professional and technical personnel involved with the preparation, distribution, and administration of TPN solutions. Such personnel include other pharmacists, nurses, and auxiliary staff handling these solutions. In addition, the pharmacist should participate in the teaching and training of undergraduate students in pharmacy, medicine, and nursing, as well as in the various continuing education programs, dealing with TPN.

VI. MONITORING THE PATIENT

The pharmacist may also have a responsible clinical role to perform in some centers. Where feasible the pharmacist should make rounds as a member of the TPN team and participate in overall patient care. The pharmacist can keep records of the patient's "profile", monitor laboratory tests and drug therapy, and be alert to problems that may arise from the administration of drugs concurrently with TPN. The pharmacist working with the physician can make recommendations in regard to solutions and electrolyte changes, and furthermore, provide alternatives in the event that incompatibilities become apparent. The pharmacist has a very useful role in advising the nurse about the safe delivery of the TPN solutions and other required medicines. In addition, he/she can offer assistance about the use of various administration sets and in-line filters.

VII. FORMULATION AND IMPLEMENTATION OF POLICIES AND PROCEDURES

The pharmacist must take responsibility for:

1. Policies and procedures for the manufacture, quality control, storage, and delivery of parenteral nutrition solution.
2. Detailed records of the formulations released and of special parenteral nutrition orders.
3. Policies and procedures for maintenance and quality control of equipment used in the preparation of TPN solutions.
4. The needs for program expansion.
5. Aiding the physician in performing clinical trials and maintaining records for this purpose.
6. Contributing to the development of protocols for the TPN program.

VII. ESTABLISHING EFFECTIVE COMMUNICATION

The pharmacist has a vital role as a communicator of information to other personnel involved directly or indirectly with the TPN program. This will include participation in the work of various committees such as the Pharmacy and Therapeutics Committee, the TPN Committee, and the Infection Control Committee as well as preparation of special bulletins or memos for nurses and medical staff. In addition the pharmacist is responsible for preparing documents outlining costs of solutions, equipment, and other pharmaceutical expenses related to TPN. Furthermore, the pharmacist must develop in cooperation with the other members of the TPN team a concise and informative TPN order form to ensure effective communication with the prescribing physician. It is imperative that clear and complete information be provided about the finished solutions so that those involved in the administration of such solutions are fully informed of their composition. The maintenance of proper records must also be considered to be an effective means of communication. Good pharmacy records can provide vital information to other dispensers. It is of the utmost importance to cooperation and effective patient care that good and effective communication develops between all members affected by the TPN program.

IX. INVOLVEMENT IN RESEARCH

The pharmacist should participate in the various types of research related to TPN. This includes new formulations, product development, and establishment of new procedures for the preparation, storage, and distribution of solutions, utilization review, and cost effectiveness of the program. There is, in addition, scientific research with formulation problems such as compatibilities and incompatibilities, packaging systems, clinical studies, record maintenance, and product evaluation, and joint research with other members of the team.

X. INVOLVEMENT WITH THE AMBULATORY PATIENT

The pharmacist has a valuable and demanding role in this expanding area of TPN. Responsibilities in this area will include:

1. Educating patients (or their families)
2. Procurement and distribution of supplies
3. Scheduling and preparing the solutions
4. Keeping up-to-date patient profiles
5. Solving problems presented by the patient
6. Making necessary formulation changes as requested by the physician
7. Communicating with other members of the TPN team regarding various problems and participating in resolving such problems
8. Checking all materials dispensed by the technician prior to their release to the patient
9. Participating in, and evaluating, new systems developed for the administration of TPN solutions
10. Preparation of a cost-accounting report for each ambulatory patient

XI. FORMULATION AND MANUFACTURE OF SPECIAL TPN-RELATED STERILE EXTEMPORANEOUS PRODUCTS

The pharmacist has training and expertise in the manufacture of sterile extemporaneous products, so that when provided with the necessary facilities and equipment he can offer most valuable assistance to the physician. Such service includes the manufacturing of:

1. Special electrolyte solutions
2. Products that are not available commercially
3. Special dosage forms

In summary, the pharmacist has unlimited opportunities to participate in the field of total parenteral nutrition. More pointedly he/she has a very responsible role as a member of the TPN team to ensure that the patient's nutritional requirements are met with safety.

READING LIST

Conrad, W. F., Gassett, R. H., and Goupil, D. A., A service-integrated approach to clinical pharmacy practice through a hyperalimentation program, *Am. J. Hosp. Pharm.*, 30, 695, 1973.

Greenlaw, C. W., Pharmacist as team leader for total parenteral nutrition therapy, *Am. J. Hosp. Pharm.*, 36, 648, 1979.

Madan, P. L., Madan, D. K., and Palumbo, J. F., Total parenteral nutrition, *Drug Intel. Clin. Pharm.*, 10, 684, 1976.

McLeod, D. C., Contribution of clinical pharmacists to patient care, *Am. J. Hosp. Pharm.*, 33, 904, 1976.

Powell, J. R.and Cupit, G. C., Developing the pharmacist's role in monitoring total parenteral nutrition, *Drug. Intell. Clin. Pharm*, 8, 576, 1974.

Schloesser, L., Hryciuk, L., and Hoffman, D. M., Parenteral nutrition — a team concept from the pharmacist's viewpoint, *Am. J. I. V. Ther.*, p. 42, 1979.

Schneider, P., The pharmacist's role in inpatient and home care hyperalimentation programs, *Hosp. Pharm.*, 13, 71, 1978.

Shaw, J. and Lamy, P. P., A total parenteral nutrition program for a community hospital, *Hosp. Formul.*, 583, 1977.

Skoutakis, V. A., Martinez, D. R., Miller, W. A., and Dobbie, R. P., Team approach to total parenteral nutrition, *Am. J. Hosp. Pharm.*, 32, 693, 1975.

Sauvé, Sister Frances, The pharmacist and a nutritional intravenous therapy program, *Am. J. Hosp. Pharm.*, 28, 106, 1971.

Tanner, D. J., Comprehensive pharmacy services in an 85-bed hospital, *Am. J. Hosp. Pharm.*, 33, 340, 1976.

Chapter 13

PRODUCTS FOR PARENTERAL NUTRITION

George Tsallas

TABLE OF CONTENTS

I. INTRODUCTION

During the past ten years a number of commercial products have become available for parenteral nutrition. Differences in composition and cost allow one to choose from amongst this variety. Commercially available solutions for parenteral nutrition can be divided into four main categories, providing sources of protein (amino acids), energy (calories), minerals (including trace elements), and vitamins.

II. PROTEIN SUPPLEMENT SOLUTIONS

A. Protein Hydrolysates

These solutions are prepared by enzymatic hydrolysis of proteins such as casein or fibrin. They contain both essential and nonessential amino acids with some peptides and electrolytes. Of the protein content, 10 to 20% is in the di- or tripeptide form. Only 50 to 55% of the total nitrogen is in the alpha-amino form, with approximately 40% in the biologically inferior peptide form. These hydrolysates were the first commercially available sources of nitrogen for intravenous use. They were inexpensive and thus were widely used. However, because of their low content of utilizable nitrogen, high peptide and ammonia content, unsuitability in patients with renal and hepatic disease and in those allergic to the protein source, they are being phased out by the newer solutions containing defined amounts of crystalline amino acids. Commercially available protein hydrolysates in North America include:

Casein hydrolysate — Prepared by enzymatic hydrolysis of casein protein from milk. Commercial products available include Amigen® 5% and dextrose 5%, and Travamin® 5% and dextrose 5% from Baxter or Travenol Laboratories (see Table 1). C.P.H. 5% and Hyprotigen® 5% are other commercial products, available from Cutter and McGraw, respectively.

Fibrin hydrolysate — Produced by acid hydrolysis of beef and pork blood fibrin protein. Commercial products include Aminosol® 5% from Abbott Laboratories (see Table 1).

B. Crystalline Amino Acid Mixtures

These products are prepared using synthetic L-amino acids, and include both essential and nonessential amino acids with varying amounts of electrolytes. Utilizable nitrogen in these products is claimed to vary between 1.12 to 1.36 g/100 m*ℓ* of solution. Compared with protein hydrolysates these solutions do not contain nonutilizable nitrogenous products, are chemically defined, and do not cause hypersensitivity reactions. Earlier amino acid mixtures contained the hydrochloride form of lysine, histidine, and arginine, which caused hyperchloremic acidosis in patients. Hence recently formulated mixtures contain the acetate form of these basic amino acids. Thus the H^+ ion liberated by the metabolism of these basic amino acids is neutralized by OH^- ion produced by metabolism of the acetate. The substantially higher cost of these solutions over that of the protein hydrolysates has restricted their wider usage. Table 2 shows the composition of several commercial products. In the light of their amino acid composition one may make the following observations:

Travasol® — Contains no aspartic acid, cystine, glutamic acid, or serine and has a large amount of alanine and glycine. Approximately 39% of total amino acids are essential and 16% are branched-chain. Uses bisulfite as preservative.

FreAmine II® — Contains no aspartic and glutamic acids and no tyrosine. Approximately 49% of total amino acids are essential and 22% are branched-chain. Uses bisulfite as preservative.

Aminosyn® — Contains no aspartic acid, cystine, or glutamic acid. Essential amino acids comprise approximately 47% of the total amino acids and branched-chain 25% of the total. Uses bisulfite as preservative.

Table 1
COMPOSITION OF COMMERCIALLY AVAILABLE PROTEIN HYDROLYSATE SOLUTIONS (CONCENTRATION PER LITER)

Amino Acid Content (g/l)	Amigen 5%	Amigen 5% Dextrose 5%	Aminosol 5%	Travamine 5% Dextrose 5%
ESSENTIAL AMINO ACIDS				
L-Leucine	4.10	4.10	6.36	4.10
L-Isoleucine	2.60	2.60	2.18	2.60
L-Lysine	3.10	3.10	4.00	3.10
L-Valine	3.10	3.10	1.63	3.10
L-Tryptophan	0.35	0.35	0.50	0.35
L-Phenylalanine	2.00	2.00	1.00	2.00
L-Threonine	1.90	1.90	2.32	1.90
L-Methionine	1.30	1.30	1.00	1.30
NON- & SEMI- ESSENTIAL AMINO ACIDS				
L-Histidine	1.30	1.30	1.16	1.30
L-Arginine	1.80	1.80	2.90	1.80
L-Alanine	1.50	1.50	2.21	1.50
L-Serine	3.00	3.00	3.40	3.00
L-Tyrosine	0.60	0.60	1.10	0.60
Amino Acetic Acid (Glycine)	1.10	1.10	2.08	1.10
L-Proline	4.50	4.50	3.15	4.50
L-Glutamic Acid	13.00	13.00	1.35	13.00
L-Cysteine	-	-	0.30	-
L-Aspartic Acid	3.50	3.50	-	3.50
TOTAL PROTEIN (g)	Casein Hydrol. 50	Casein Hydrol. 50	Fibrin Hydrol. 50	Casein Hydrol. 50
Approximate PROTEIN EQUIVALENT (g)	40	40	42	40
Approximate TOTAL NITROGEN (g)	6.5	6.5	6.5 - 7.0	6.5
ALPHA-AMINO NITROGEN	55-60% of Total N	55-60% of Total N	55% of Total N	55-66% of Total N
ELECTROLYTES (mEq)				
Sodium (Na^+)	35	35	10	35
Potassium (K^+)	19	19	17	19
Calcium (Ca^{2+})	5	5	-	5
Magnesium (Mg^{2+})	2	2	-	2
Chloride (Cl^-)	20	20	7	20
Phosphate (HPO_4^{2-})	30	30	-	30
STABILIZERS (g)				
a) Sodium Bisulfite	0.5	0.5	-	0.5
b) Potassium Metabisulfite	0.56	0.56	0.6	0.56
OSMOLARITY (mOsm/l)	430	682	292	682
pH	5.0 - 7.0	5.0 - 7.0	5.0 - 6.0	5.0 - 7.0
APPROXIMATE CALORIES	170	340	175	340
SUPPLIER	BAXTER	BAXTER	ABBOT	BAXTER

Table 2
COMPOSITION OF COMMERCIALLY AVAILABLE CRYSTALLINE AMINO ACID SOLUTIONS[a] (CONCENTRATION EXPRESSED PER 100 Mℓ)

Components	Travasol 5.5%	Travasol 8.5%	Travasol 10%	Travasol E.F. 5.5%	Travasol E.F. 8.5%	Travasol E.F. 10%	FreAmine III 8.5%	FreAmine III 8.5%	Nephramine	Aminosyn 3.5%	Aminosyn 5%	Aminosyn 7%	Aminosyn 10%	Vamin 7%	Vamin 7%	Veinamine 8%[b]
Essential Amino Acids (g)																
L-Isoleucine	0.263	0.406	0.480	0.263	0.408	0.480	0.590	0.590	0.560	0.252	0.360	0.510	0.720	0.390	0.390	0.493
L-Leucine	0.340	0.526	0.620	0.340	0.526	0.620	0.770	0.770	0.880	0.329	0.470	0.660	0.940	0.525	0.525	0.347
L-Lysine	0.318	0.492	0.580	0.318	0.492	0.580	0.870[i]	0.870[i]	0.640	0.252	0.360	0.510	0.720	0.385	0.385	0.667
L-Methionine	0.318	0.492	0.580	0.318	0.492	0.580	0.450[j]	0.450[k]	0.880	0.140	0.200	0.280	0.400	0.190	0.190	0.427
L-Phenylalanine	0.340	0.526	0.620	0.340	0.526	0.620	0.480	0.480	0.880	0.154	0.220	0.310	0.440	0.545	0.545	0.405
L-Threonine	0.230	0.356	0.420	0.730	0.356	0.420	0.340	0.340	0.400	0.182	0.260	0.370	0.520	0.300	0.300	0.160
L-Tryptophan	0.099	0.152	0.180	0.099	0.152	0.180	0.130	0.130	0.200	0.056	0.080	0.120	0.160	0.100	0.100	0.080
L-Valine	0.252	0.390	0.460	0.252	0.390	0.460	0.560	0.560	0.650	0.280	0.400	0.560	0.800	0.425	0.425	0.253
Non- & Semi-Essential A.A. (g)																
L-Alanine	1.149	1.760	2.080	1.149	1.760	2.080	0.600	0.600	-	0.448	0.640	0.900	1.280	0.300	0.300	-
L-Arginine	0.570	0.880	1.040	0.570	0.880	1.040	0.310	0.810	-	0.343	0.490	0.690	0.980	0.300	0.300	0.749
L-Histidine	0.241	0.372	0.400	0.241	0.372	0.440	0.240	0.240	-	0.105	0.150	0.210	0.300	0.240	0.240	0.237
L-Proline	0.230	0.356	0.420	0.230	0.356	0.420	0.950	0.950	-	0.300	0.430	0.610	0.860	0.810	0.810	0.107
L-Serine	-	-	-	-	-	-	0.500	0.500	-	0.147	0.210	0.300	0.420	0.750	0.750	-
L-Tyrosine	0.022	0.034	0.040	0.022	0.034	0.040	-	-	-	0.031	0.044	0.044	0.044	0.050	0.050	-
Amino Acetic Acid (Glycine)	1.140	1.760	2.080	1.140	1.760	2.080	1.700	1.190	-	0.448	0.640	0.900	1.280	0.210	0.210	3.387
L-Cysteine	-	-	-	-	-	-	0.020	0.020	-	-	-	-	-	0.140	0.140	-
L-Aspartic Acid	-	-	-	-	-	-	-		-	-	-	-	-	0.405	0.405	0.400
L-Glutamic Acid	-	-	-	-	-	-	-		-	-	-	-	-	0.900	0.900	0.426
Total Amino Acids (g)	5.500	8.500	10.000	5.500	8.500	10.000	8.500	8.500	5.000	4.500	5.000	7.000	9.964	7.000	7.000	8.000
Total Essential Amino Acids (g)	2.160	3.340	3.940	2.160	3.340	3.940	4.190	4.190	5.090	1.645	2.350	3.320	4.700	2.860	2.860	2.827
Total Aromatic AA.	0.439	0.678	0.800	0.439	0.678	0.800	0.610	0.610	1.080	0.210	0.300	0.430	0.600	0.645	0.645	0.480
Total Branched-Chain AA. (g)	0.855	1.322	1.560	0.855	1.322	1.560	1.920	1.920	2.090	0.861	1.230	1.730	2.460	1.540	1.540	1.093
Total Nitrogen (g)[c]	0.924	1.420	1.680	0.924	1.420	1.680	1.250	1.300	0.587	0.550	0.786	1.100	1.569	0.940	0.940	1.330
Alpha-Amino Nitrogen (g)[d]	0.639	1.088	1.265	0.639	1.088	1.629	1.057	1.002	0.507	0.414	0.593	0.528	1.186	0.761	0.761	1.086
Percent Nitrogen[e]	6.8	16.71	6.8	16.8	16.8	16.95	14.71	15.30	11.74	15.71	15.72	15.71	15.69	13.43	13.43	16.63
Amino Acid/Nitrogen Ratio[f]	5.956	5.987	5.977	5.956	5.987	5.977	6.808	6.538	8.671	6.304	6.303	1.026	6.289	7.410	7.410	5.593
Protein Equivalent[g]	5.775	8.875	10.500	5.775	8.875	10.500	7.813	8.125	3.669	3.438	4.913	6.875	9.806	5.875	5.875	8.313
E/T Ratio[h]	2.338	2.352	2.345	2.338	2.345	3.357	8.671	3.223	2.991	2.990	2.990	3.018	2.996	3.043	3.043	2.126
Electrolytes (mEq)																
Sodium (Na+)	7.0	7.0	7.0	-	-	-	1.0	1.0	-	4.00	-	-	-	5	5	4
Potassium (K+)	6.0	6.0	6.0	-	-	-	-	-	-	1.84	0.54	0.54	0.54	2	2	3
Magnesium (Mg^{2+})	1.0	1.0	1.0	-	-	-	-	-	-	0.30	-	-	-	0.3	0.3	0.6
Calcium (Ca^{2+})	-	-	-	-	-	-	-	-	-	-	-	-	-	0.5	0.5	-
Chloride (Cl^-)	7.0	7.0	7.0	2.2	3.4	4.0	-	0.2	-	4.0	-	-	-	5.5	5.5	5
Acetate	10.0	13.5	5.0	3.5	5.2	8.7	4.2	7.4	-	2.7	3.0	3.0	3.0	-	-	5
Phosphate	6.0	6.0	6.0	-	-	-	2.0	2.0		-	-	-	-	-	-	-
Calculated Carbohydrate (kcal)	-	-	-	-	-	-	-	-		-	-	-	-	37	-	-
Stabilizer					-	-										
1) Sod Bisulfite (mEq)	0.3	0.3	0.3	0.3	0.3	0.3	0.1(g)	0.1(g)	-	-	-	-	-	-	-	-
2) Pot. Metabisulfite (gm)	-	-	-	-	-	-	-		0.05	0.06	0.06	0.06	0.06	-	-	0.01
Approximate pH	6.0	6.0	6.0			6.0	6.5	6.5	6.0	5.3	5.3	5.3	5.3	5.2		6.2-6.5
Calculated Osmolarity (mOsm/l)	860	1160	1300	520	880		840	810	420	480	500	700	979	1115	690	950
Unit Size (ml)	500	500	250,100	500	500	1000	500	500	500	1000	250,500,1000	500	500	500	500	500
Supplier	BAXTER						ABBOTT							PHARMACIA		CUTTER

a From manufacturer's monograph
b Not available in Canada
c Calculated from percent of nitrogen per amino acid
d Alpha-amino nitrogen content of amino acids
e Amount of nitrogen per 100 g of amino acids
f It is: $\frac{\text{Total amino acids (g)}}{\text{Total nitrogen (g)}}$
g Obtained using relationship: 1 g of nitrogen is equivalent to 6.25 g of protein. Therefore protein equivalent = total nitrogen X 6.25
h $E/T = \frac{\text{Total essential amino acids (g)}}{\text{Total nitrogen (g)}}$
i As L-Lysine Acetate (0.620 g free base)
j As DL-Methionine
k As D-Methionine

Nephramine® — Contains 100% essential amino acids of which approximately 42% is of the branched-chain category. Uses bisulfite as preservative.
Vamine N® or F® — Contains all listed essential and nonessential amino acids. Essentials and branched-chain comprise 47% and 25% of the total, respectively. This product is the only amino acid solution with no bisulfite preservative.
VeinAmine® — Contains no serine, tyrosine, or cystine. Essential amino acids comprise 35% of the total amino acids and branched-chain amino acids 13% of total. Uses bisulfite as preservative.

The presence of bisulfite as a preservative in most of these solutions has been implicated as the cause of abnormally elevated SGOT levels (Grant et al.) and is an area that should be investigated.

The effect of these amino acid mixtures, including Travasol®, Vamine N®, and Aminosyn® on nitrogen balance and plasma amino acid levels was studied by Phillips and Odgers using critically ill patients without liver or renal failure.

They found that plasma amino acid concentrations could vary in a manner directly explicable by the composition of the solution administered; thus the plasma level of glycine was higher with Travasol® and of phenylalanine higher with Vamin N®. In contrast to this rule plasma concentrations of cystine, glutamic acid, serine, and tyrosine were normal, even when these amino acids were absent from the infusion.

Urinary excretion of amino acids and nitrogen balance depended more on nitrogen and energy input, relative to the degree of catabolism of the patient, than on the composition of the infused solution.

Although we have many amino acid solutions available the ideal one has yet to be developed. With better understanding of nutritional therapy, different types of solutions may eventually be made available for different ages and disease states. At present however, there is conflicting evidence on the optimum amino acid composition of these solutions. Thus selection of amino acid solutions is based primarily on cost, supply, and convenience.

III. ENERGY (CALORIC) SOURCE SOLUTIONS

A. Carbohydrates

There are many carbohydrate sources capable of being utilized for energy. However, only a few can be physically dissolved and infused parenterally. These include dextrose, fructose, sorbitol, and xylitol. Of these sources only dextrose is completely nontoxic and utilized by the human body even at high rates of infusion. Dextrose is the cheapest available energy source, providing 3.4 to 4.0 kcal/g. It is commercially available as a sterile injection of various concentrations ranging from 5 to 70% w/v.

B. Fat

This source of energy is available as a stable sterile fat emulsion composed of either soybean or safflower oil, emulsified with purified egg lecithin phosphatides. Fat is a high-density caloric source providing 9.1 cal/g of fat. In addition, fat will provide essential fatty acids such as linoleic and linolenic that we cannot manufacture in our body. The fat emulsion is isotonic even at concentrations of 30% and thus can be administered alone or with amino acids, carbohydrates, and other nutrients into a peripheral vein.

As a 10% emulsion its caloric density is comparable to hypertonic dextrose, but as a 20% emulsion it provides twice the caloric density of hypertonic dextrose allowing the infusion of energy in a smaller volume. This is advantageous in patients with cardiac and renal failure where the volume of fluid given may need to be restricted.

Table 3
COMPOSITION OF FAT EMULSIONS AVAILABLE IN NORTH AMERICA

Components[a]	Nutralipid® 10% (Intralipid®)	Nutralipid® 20% (Intralipid®)	Liposyn® 10%	Liposyn® 20%
Soybean oil	10	20		
Safflower oil			10	20
Egg phosphatide	1.2	1.2	1.2	1.2
Glycerin	2.5	2.5	2.5	2.5
Caloric content/mℓ	1.1	2.0	1.1	2.0
Fatty acids (g/100 g total fatty acids)	54.3	54.3	77	77
Linoleic				
Linolenic	7.8	7.8	0.5	0.5
Oleic	26.4	26.4	13	13
Palmitic	9.2	9.2	7	7
Stearic	2.9	2.5	2.5	2.5
Other	1.2	1.2	Trace of myristic	Trace of myristic
Cholesterol (g/ℓ)	~0.045	~0.045	0.001	Not known
Tocopherols (μg/ℓ)	10	20	20 as D-α-tocopherol	
Vitamin E activity (IU/100 mℓ)	~3—3.5	~6—7	Not known	Not known
Minerals (conc. mg/ℓ)				
Sodium[b]	<20	0	<20	<20
Potassium[b]	<10	<10	<10	<10
Magnesium[b]	<0.1	<0.1	<0.1	<0.1
Calcium[b]	<0.7	<0.7	<0.7	<0.7
Phosphorus[b]	420—460	420—460	423—432	423—432
Copper	<0.05	<0.05	<0.05	<0.05
Zinc	<0.05	<0.5	<0.05	<0.5
Manganese	<0.1	<0.1	<0.1	<0.1
Iron	<0.1	<0.1	<0.1	<0.1
Chromium	<0.1	<0.1	<0.1	<0.1
Cobalt	<0.5	<0.5	<0.5	<0.5
Selenium	<0.025	<0.025	<0.025	<0.025
Osmolarity (mOsm/ℓ)	~280	~300	~300	~340
pH	~8.0	~8.0	~8.0	~8.3

[a] g/dℓ except where specified
[b] Approximate values.

There are presently two commercially available fat emulsions in North America. One of these, a soya bean oil product, is Nutralipid® (called Intralipid® in Europe where it is manufactured by Vitrum) which is distributed by Pharmacia in Canada and Cutter in both Canada and the U.S. It is the most widely used fat preparation in both North America and Europe, with a reputation for being safe and efficacious. Clinical experience with it now spans well over ten years. The other commercial product is a more recent one composed of safflower oil. It is marketed by Abbott Laboratories under the trade name Liposyn®. Both of these fat emulsions have similar emulsifying agents and additives, the major difference is in the composition of triglyceride used as the caloric source. A summary of the components is given in Table 3.

Both of these products are available in concentrations of 10% and 20% with a caloric density of 1.1 and 2.0 kcal/mℓ, respectively. Both fat emulsions have been approved for storage at temperatures not greater than 25°C.

At some centers (notably that of Solassol in France) fat emulsions have been premixed prior to infusion with amino acid, carbohydrate, and electrolyte solutions. However, this practice is not recommended until controlled studies have been done to show stability under such conditions. At present it is advocated that fat emulsions be administered with other nutrients only by the use of a Y connector near the infusion site just prior to entry into the patient. There is some evidence that premixing the fat emulsion with the amino acid, dextrose, and electrolyte solution in the same container may result in the development of a gummy precipitate which may clog the catheter (Messing — personal communication). In-line filters smaller than 14 μm should not be used to filter fat emulsions. Such filters will withhold the emulsion globules.

Although Frank and Brian have demonstrated compatibility of Nutralipid® (Intralipid®) with a number of medications, with the exception of heparin sodium drugs should not be added to the emulsion. Details of infusion rates of these fat emulsions are discussed in Chapter 3.

The most commonly noted side effects with these products are nausea, the taste of the emulsion, and vomiting. Diarrhea may occur in some patients (Connon et al.).

These commercial products are packaged in bottles of 500, 200, and 100 mℓ. They are relatively expensive (i.e., $20 to $40/500 mℓ). The containers are single-dose units and any partially used bottles must be discarded after one infusion. It is recommended that bottles should not be hung longer than 12 hr during administration because these emulsions readily support bacterial growth.

IV. MINERALS

A. Electrolytes

This group of minerals includes salts of sodium, potassium, magnesium, calcium, and phosphorus. These are presently commercially available as single salts or mixtures for injection. Single salts available for injection include:

- Sodium chloride, acetate, lactate, or bicarbonate
- Potassium chloride, acetate, lactate, or phosphate
- Calcium chloride, gluconate, or gluheptate
- Magnesium chloride or sulfate
- Phosphate, usually as sodium or potassium phosphate

Multiple-component electrolyte injections contain admixtures of sodium, potassium, calcium, and magnesium as chloride and acetate salts. These solutions are buffered to chemically stabilize the component electrolytes. They are usually available as multidose containers of 30 to 50 mℓ. Multiple-component electrolyte injections have an advantage in that when used in making up TPN mixtures fewer additions have to be made, thus saving time and cost.

Some commercially available multiple-electrolyte parenteral preparations are shown in Table 4.

Where manufacturing facilities and suitable personnel are available electrolyte injections may be economically manufactured locally. Under these circumstances they can be packaged in an appropriate unit size to meet the needs of the locally devised TPN protocol.

Table 5 shows electrolyte injectables extemporaneously manufactured at the Toronto General Hospital pharmacy department. Manufacturing instructions, including equipment, are available from the authors.

The packaging of these products in containers of a larger unit size has been found not only to be economical, but also practical for bulk compounding of TPN solutions.

Table 4
COMMERCIALLY AVAILABLE MULTIPLE-ELECTROLYTE ADDITIVE SOLUTIONS

	Hyperlyte® (McGaw)		Multiple Electrolyte Additive (IMS)		TPN Electrolytes (Abbott)	
Electrolyte	Ion conc. (meq/25 mℓ)	Salt	Ion conc. (meq/mℓ)	Salt	Ion conc. (meq/20 mℓ)	Salt
Sodium (Na^+)	25	Sodium acetate · $3H_2O$ Sodium gluconate	0.8	Sodium chloride	35	Sodium chloride Sodium acetate
Potassium (K^+)	40.5	Potassium chloride	0.4	Potassium chloride	20	Potassium chloride
Calcium (Ca^{2+})	5	Calcium acetate · H_2O	0.096	Calcium gluconate	4.5	Calcium chloride · $2H_2O$
Magnesium (Mg^{2+})	8	Magnesium acetate · $4H_2O$	0.16	Magnesium sulfate · $7H_2O$	5.0	Magnesium chloride · $6H_2O$
Acetate[a] (CH_3COO^-)	40.6				29.5	
Gluconate ($CH_2(OH) \cdot (CH \cdot OH)_4 \cdot COO^-$)	5.0		0.096			
Chloride (Cl^-)	33.5		1.2		35	

[a] Includes acetic acid used to adjust pH.

Table 5
STERILE ELECTROLYTE TPN ADDITIVE SOLUTIONS[a]

Electrolyte	Chemical source	Supplier of chemical	Chemical grade	Quantity of chemical (g/mℓ)	Electrolyte ion conc. (meq/mℓ)	Unit size (mℓ)	Cost per unit ($)
Sodium (Na^+)	Sodium chloride	Anachemia	Analytical or USP	0.2337	4	250	2.40
Potassium (K^+)	Potassium chloride	Anachemia	Analytical or USP	0.2982	4	250	2.30
Magnesium (Mg^{2+})	Magnesium chloride · $6H_2O$	British Drug House	USP	0.4068	4	250/30	4.65/1.25
	Magnesium sulfate · $7H_2O$	Fisher	Certified reagent	0.5000		250	3.10
Calcium (Ca^{2+})	Calcium chloride · $2H_2O$	Fisher	Certified Reagent	0.2940	4.50	250	4
	Calcium gluconate	British Drug House	USP	0.0927	0.45	250	3.20
Phosphate (PO_4^{2-})	Potassium phosphate (*mono- and dibasic*)	Fisher	Certified reagent	0.1635 (mono), 0.1570 (di)	2.0 mmol (65 mg P/mℓ)	150/30	2.65/0.75
Lactate ($CH_3 \cdot CHOH \cdot COO^-$)	Sodium lactate	British Drug House	Certified reagent	[b]	4	250	3.60
Sodium (Na^+)	Sodium chloride Sodium acetate · $3H_2O$	Anachemia	USP or certified reagent	0.0731 0.1360	2.25	250	2.55
Potassium (K^+)	Potassium chloride	British Drug Houses	USP	0.0746	1.0		
Calcium (Ca^{2+})	Calcium chloride · $2H_2O$	Fisher	Certified reagent	0.0184	0.25		
Magnesium (Mg^{2+})	Magnesium chloride · $6H_2O$	British Drug Houses	USP	0.0407	0.4		
Chloride (Cl^-) Acetate (CH_3COO^-)					2.9 1.04		

[a] As manufactured by the Department of Pharmaceutical Services, Toronto General Hospital, Canada.
[b] Prepared by neutralizing sodium hydroxide with lactic acid.

V. VITAMINS

Both fat-soluble and water-soluble vitamins are available as sterile aqueous injections for addition to the TPN admixture. There are many commercially available parenteral multivitamin preparations that can be added to the TPN solution. Some of the most common commercially available parenteral products are shown in Table 6. These products are usually added to the TPN mixture just prior to its infusion.

VI. TRACE ELEMENTS

Micronutrients currently found to be essential for man include copper, zinc, chromium, selenium, iodine, iron, molybdenum, and cobalt (as Vitamin B_{12}). Other trace elements such as manganese, nickel, silicon, tin, and vanadium while essential in animals have not yet been proven to be essential for man.

A. Availability

1. Blood or Plasma

Trace elements have been made available to patients by periodic infusions of blood or plasma (Table 7). From this table it is obvious that unless 1 to 2 ℓ of plasma is infused per day the requirements as defined in Chapter 3 will not be met.

Hence this method of providing trace element needs is of questionable efficacy, is expensive, incompatible with TPN solutions, and bears the risk of transmitting hepatitis.

2. Commercial Parenteral Products

Trace elements are present as contaminants in various commercial intravenous fluids and intravenous additives. Amounts present vary with the type of solution and lot number. The most commonly reported trace elements in commercial parenteral fluids are zinc, copper, and manganese. These elements are introduced into the parenteral fluids unintentionally during the manufacturing process. Since trace element analysis is not required or performed routinely during manufacture these elements are not identified on the label. Packaging components such as rubber closures are good sources of zinc because zinc salts are employed in the vulcanization process of the rubber closure. From the rubber stopper it is released into the packaged parenterals during storage or steam sterilization. Because of the accidental nature of the introduction of these elements into parenteral solutions, they cannot be relied upon to provide consistent and uniform levels of trace elements. It is important, however, that the approximate range of contamination be recognized. Tables 8 to 12 summarize currently available data on the trace element content of various commercially available intravenous fluids commonly used to make TPN admixtures.

3. Sterile Injectables

To avoid the inherent disadvantages of supplying trace elements by blood and products derived from blood, and also to overcome the capricious nature of the trace element content of the commercial solutions, it is necessary to add defined amounts of sterile trace-element solution to the TPN admixture. The preparation of these solutions is the responsibility of hospital pharmacists. These solutions are becoming available commercially.

There are basically two types of trace element formulations, (1) "cocktails" or multiple trace-element solutions, and (2) single-entity solutions.

B. Multiple Trace-Element Solutions

In general, formulations reported to date in the literature contain multiple elements.

Table 6
COMPOSITION OF SOME COMMERCIALLY AVAILABLE MULTIVITAMIN INJECTIONS COMMONLY USED IN TOTAL PARENTERAL NUTRITION

Contents and units	MVI-1000[a] injection	MVI-1000[b] injection	MVI-Concentrate[b]	MVI-12[a,b]	VI-Syneral injection[b]	Pancebrin injection[b]	Solu-Zyme with ascorbic acid[a]	Folbesyn vit. B&C[b]	Folbesyn (content /vial)[a]	Bejex [b]	Vi-Cert C 1000 with B_{12}[b]	LyoB-C with B_{12}[b]	Vi-Cert C-500 (content /vial)[a]	Beminal with C Fortis (content /vial)[a]	Berocca-C[b,c]
Vitamin A, IU	1000	1000	2000	660	5000	5000	—	—	—	—	—	—	—	—	—
Vitamin D, IU	100	100	200	40	500	500	—	—	—	—	—	—	—	—	—
Vitamin E, IU	1.0	0.5	1.0	2.0	1.0	1.1	—	—	—	—	—	—	—	—	—
Thiamine (B_1) mg	4.5	5.0	10.0	0.6	5.0	5.0	2.0	5.0	10.0	2.0	25.0	14.3	25.0	50.0	5.0
Riboflavin (B_2) mg	1.0	1.0	2.0	0.72	0.5	1.0	2.0	5.0	10.0	2.0	10.0	1.4	10.0	5.0	5.0
Niacinamide, mg	10.0	10.0	20.0	8.0	10.0	10.0	50.0	37.0	75.0	50.0	100.0	35.7	100.0	125.0	40.00
Pantothenic acid or derivative, mg	2.6	2.5	5.0	3.0	2.5	1.5	9.0	5.0	10.0	10.0	20.0	7.1	20.0	10.68	10.5
Pyridoxine (B_6), mg	1.2	1.5	3.0	0.8	1.5	1.5	1.0	7.5	15.0	1.0	20.0	1.4	20.0	5.0	10.00
Cyanocobalamin (vit. B_{12}), μg	—	—	—	1.0	—	—	5.0	7.5	15.0	5.0	25.0	3.6	—	—	—
Folic acid, mg	—	—	—	0.08	—	—	1.0	1.5	—	—	—	—	—	—	—
Ascorbic acid (vit. C), mg	100	—	100	20.0	25.0	30.0	100	150	300	100	1000	28.6	500	500	50.0
D-Biotin, μg				12.0											100

Note: Content is per mℓ unless otherwise stated.

[a] Product available in Canada.
[b] Product available in the U.S.
[c] Product available in Canada upon special request.

From *Product Monographs and Facts and Comparisons*, Kastrup, Irwin, K., Ed., Gene H. Schwach, St. Louis, Mo., 1981. With permission.

Table 7
TRACE ELEMENTS IN BLOOD AND PLASMA

Trace element	Whole blood (μg/ℓ)	Plasma (μg/ℓ)
Co	0.35	0.29
Cr		(27—180)
Cu	(720—1060)	(980—1200)
Fe	(440—560 mg)	(580—2100)
Mn	200	(1.8—3.1)
Se	(20—220)	
Zn	(8.84—16.0 mg)	(1.1—3.0 mg)
F	(40—360)	(140—190)
I		(38—60)

From Kartinos, N. J., in *Advances in Parenteral Nutrition*, Johnston, I. D. A., Ed., University Park Press, Baltimore, 1978, 233. With permission.

Although these formulations apparently simplify manufacturing and clinical use, they have the disadvantage of having a fixed ratio of trace elements. Thus the amount of trace elements given cannot be tailored to the patient's needs. In addition one must employ the halogen salts of the elements for maximum stability of the solution. Controlled studies have shown that this restriction is undesirable as needs of copper and zinc may vary quite independently of each other — see Chapter 3.

Various multiple-component formulations are presented in Table 13. Details of the preparation of these formulations may be obtained from the cited references.

C. Single-Entity Trace Element Solutions

More recently the A.M.A. Nutrition Advisory Group has recommended that trace elements be available in single-component solutions. The reasons behind this recommendation are

1. Requirements of individuals for each trace element vary with age, clinical, and metabolic status.
2. Precipitation has been reported when multiple trace elements are mixed together.
3. Due to variability in requirements, toxicity from some trace element may result from overdosage of one element even though there may be a concurrent deficiency of others.

Single-entity trace-element injections recently have become available commercially (see Table 14). However, cost, lack of availability in certain countries, and more importantly, specific institutional protocols, still make it necessary for various pharmacies to manufacture their own solutions.

D. Extemporaneous Manufacture of Trace Elements

Extemporaneous manufacturing of trace elements by a hospital pharmacy requires adequate facilities and equipment suitable for the compounding, packaging, and sterilizing of the product. In addition it requires specially trained and knowledgeable personnel in the areas of sterile product formulation.

E. Manufacturing Technique

Both multiple-component and single-entity solutions are compounded using high-purity analytical grade trace element salts. The manufacturing technique may vary with

Table 8
TRACE-ELEMENT CONTENT OF COMMERCIALLY AVAILABLE AMINO ACID SOLUTIONS

Solution	Manufacturer	Cu (mg/ℓ)	Zn (mg/ℓ)	Mn (mg/ℓ)	Cr (mg/ℓ)	Se (μg/ℓ)	I (mg/ℓ)	Fe (mg/ℓ)	Ref.
Travasol® 10%	Baxter	0.05	0.7—2.7	0.05	0.05	10—13	5—5	0.1	Pipa et al.
Travasol® 10% E.F.	Baxter	0.05	0.7—3.2	0.05	0.05	7—10	5	0.1	Pipa et al.
Travasol® 8.5%	Baxter	0.05	0.10	0.07	0.37	10	—	0.1	Pipa et al.
Travasol® 8.5% E.F.	Baxter	0.05	0.10	0.05—0.10	0.25—0.50	10	—	0.1	Pipa et al.
Travasol® 5.5%	Baxter	0.05	0.12	0.05	0.5	10	—	0.2	Pipa et al.
Travasol® 5.5% E.F.	Baxter	0.05	0.07	0.05	0.5	10	—	0.1—0.2	Pipa et al.
Vamin N® 7%	Pharmacia	0.05—0.1	1.1—1.6	0.05	0.05	5	5	0.1	Pipa et al.
	Vitrum	0.3—0.4	2.9—9.8			0.01	5	0.1	van Caillie et al.
Vamin F® 7%	Pharmacia	0.05	0.5	0.05	0.05	5	5	0.2	Pipa et al.
	Vitrum	0.3—0.4	3.9—4.3						
FreAmine® II	McGaw	0.05	3.8—5.3	0.05	0.05	36	5	0.1	Pipa et al
		0.085	1.85						Hoffman and Ashby
		0.041—0.065	0.558—1.88	0.059—0.172					Shearer and Bozian
						32			Smith and Goos
		ND	1.7—3.1	0.0033					Shils and Alcock
		0.009—0.011	0.8—4.04		0.0022—0.0024				Haver and Kaminski
Aminosyn® 5%	Abbott	0.05	2.6	0.05	0.05	18	5	0.4—0.6	Pipa et al.
		0.008	0.075		0.0008			0.038	Haver and Kaminski
		0.004	0.289—0.327		0.01—0.014				Carlson et al.
Aminosyn® 7%	Abbott	0.05	1.6	0.05	0.05	18	5	0.5	Pipa et al.
		0.005—0.007	0.223—0.469		0.012—0.019				Pipa et al.
		ND	0.12—0.14						Haver and Kaminski
Aminosyn® 10%	Abbott	0.05	2.35	0.05	0.05	20	5	0.2	Pipa et al.
		ND	0.03—0.06						Haver and Kaminski
Amigen® 5%	Baxter	0.05	5.8	0.25	0.05	27	5	1.5	Pipa et al.
		0.02	1.19					1.16	Haver and Kaminski
		0.04—0.06	2.83—4.01	0.209—0.266					
		0.078	1.438						Hoffman and Ashby

Note: ND = not done.

Table 9
TRACE-ELEMENT CONTENT OF COMMERCIALLY AVAILABLE DEXTROSE SOLUTIONS

Solution	Manufacturer	Cu (mg/ℓ)	Zn (mg/ℓ)	Mn (mg/ℓ)	Cr (mg/ℓ)	Se (μg/ℓ)	I (mg/ℓ)	Fe (mg/ℓ)	Ref.
Dextrose 10%	Winthrop	0.85	2.9	0.05	0.05	5	5	0.1	Pipa et al.
	Cutter	0.01	0.025—0.158					0.1	Haver and Kaminski
	Abbott	0.0006	Not done						Shils and Alcock
Dextrose 13.3%	Baxter (Viaflex® bag)	0.05	0.65	0.05	0.05	5	5	0.2	Pipa et al.
Dextrose 33.3%	Baxter (Viaflex®)	0.05	1.3	0.05	0.05	5	5	0.2	Pipa et al.
Dextrose 20%	Travenol					273			Smith and Goos
Dextrose 50%	Cutter	0.05	0.62	0.05	0.05	5	5	0.1	Pipa et al.
		0.01	0.025—0.158						Haver and Kaminski
		0.08	Not done						Hoffman and Ashby
	Abbott	0.10	*1.3—1.95*	0.05	0.05	5	5	0.1	Pipa et al.
						470			Smith and Goos
		0.01	0.025—0.029	0.0007—0.0027				0.10	Haver and Kaminski
	Baxter (Viaflex®)	0.05	0.6	0.05	0.05	5	5	0.2	Pipa et al.
	McGaw	0.05	1.75	0.05	0.05	5	5	0.2	Pipa et al.
		0.01—0.012	0.025—0.188		0.0018			0.1	Haver and Kaminski
		0.08	Not done						Hoffman and Ashby
	Travenol					271			Smith and Goos
Dextrose 70%	McGaw	0.05	0.5	0.05	0.05	5	5	5	Pipa et al.

Table 10
TRACE-ELEMENT CONTENT IN COMMERCIALLY AVAILABLE TPN ADDITIVE SOLUTIONS

Solution	Manufacturer	Cu (mg/ℓ)	Zn (mg/ℓ)	Mn (mg/ℓ)	Cr (mg/ℓ)	Se (μg/ℓ)	I (mg/ℓ)	Fe (mg/ℓ)	Ref.
Calcium	Glaxo	0.15—0.2	0.08—1.85	0.08—0.18	0.15—0.30	<5	<5	0.5—0.6	Pipa et al.
gluconate 10%	Parke-Davis					369			Smith and Goos
Potassium	Abbott	0.2—0.25	0.3—1.7	0.1—0.2	0.3—0.7	<5	<5	1.2—1.4	Pipa et al.
chloride	Linson	0.158	0.45						Hoffman and Ashby
20 meq/10 mℓ									
40 meq/20 mℓ	Linson	0.2	1.4	0.2	0.4	<5	<5	1.2	Pipa et al.
40 meq/20 mℓ	Abbott	0.16	0.115						Hoffman and Ashby
Sodium chloride	Linson	<0.05	0.02	<0.05	<0.05	<5	<5	<0.1	Pipa et al.
0.9%		0.016	0.2—0.42						Jetton et al.
Magnesium	Sterilab	0.6	1.9	0.60	1.5	<5	<5	1.8	Pipa et al.
Sulfate 50%	Parke-Davis	0.7	0.6	0.65	1.4	<5	<5	1.7	Pipa et al.
	Elkins-Sinn	0.172	0.110						Hoffman and Ashby
Hyperlyte	McGaw	0.40	3.4	0.45	<0.05	<5	<5	1.7	Pipa et al.
Beminal with C Fortis®	Ayerst	0.05—0.1	7.6—27.1	0.28—0.38	<0.05—0.2	<5	<5	0.2—0.4	Pipa et al.
Synkavite® 10 mg/mℓ	Roche	0.05—0.1	0.05	<0.05	0.1—0.15	32—38	<5—22.5	0.1—0.2	Pipa et al.
Solu-Zyme® with Ascorbic Acid	Upjohn	0.10	4.55	0.05	0.30	<5	<5	0.4	Pipa et al.
MVI-1000®	USV	0.05	6.3—7.7	0.05	<0.05—0.4	13	<5	0.4	Pipa et al.
		0.078	2.30						Hoffman and Ashby
						529			Smith and Goos
Ascorbic acid inj. 500 mg/2 mℓ	Elkins-Sinn	0.152	0.048						Hoffman and Ashby
Folvite® (5 mg/mℓ)	Lederle	0.127	6.225						Hoffman and Ashby
Potassium phosphate (60 meq/30 mℓ)	McGaw	0.317	0.34						Hoffman and Ashby
Tagamet® 300 mg/2 mℓ	SKF	0.10	1.0—2.1	<0.05	0.35—0.40	<5	<5	0.6	Pipa et al.
Lasix® 10 mg/mℓ	Hoechst	<0.05—0.05	0.58—0.95	<0.05	0.05—0.1	<5	<5	0.1—0.2	Pipa et al.
Heparin sodium 1,000 IU/mℓ	Harris	0.05	0.3—0.65	<0.05	<0.05	<5	<5	<0.1	Pipa et al.
10,000 IU/mℓ		0.1—0.15	0.93—1.3	<0.05—0.2	<0.05—0.2	<5	<5	0.1—0.2	Pipa et al.
Toronto insulin	Connaught	0.10	19.9	0.05	<0.05	<5	<5	<0.1	Pipa et al.

Table 11
TRACE-ELEMENT CONTENT OF COMMERCIALLY AVAILABLE STERILE WATER FOR INJECTION

Solution	Manufacturer	Cu (mg/ℓ)	Zn (mg/ℓ)	Mn (mg/ℓ)	Cr (mg/ℓ)	Se (μg/ℓ)	I (mg/ℓ)	Fe (mg/ℓ)	Ref.
SWFI	Baxter (Viaflex®)	<0.05	0.75	<0.05	<0.05	<5	5	0.2	Pipa et al.
	Baxter (Bottle)	<0.05	0—0.1 0.1	<0.05	<0.05	<5	5	<0.1	Pipa et al. Jetton et al.

Table 12
TRACE-ELEMENT CONCENTRATION IN COMMERCIALLY AVAILABLE FAT EMULSIONS

Fat emulsion	Manufacturer	Cu (mg/ℓ)	Zn (mg/ℓ)	Mn (mg/ℓ)	Cr (mg/ℓ)	Se (μg/ℓ)	I (mg/ℓ)	Fe (mg/ℓ)	Ref.
Nutralipid® 10%	Pharmacia	0.05	0.5	0.1	0.1	25		0.1	Pipa et al.
Nutralipid® 20%	Pharmacia	0.05	0.5	0.1	0.1	25		0.1	Pipa et al.
Intralipid® 10%	Cutter	0.02—0.028	0.033—0.046		0.0009			0.1	Haver and Kaminski
Liposyn® 10%	Abbott	0.05	0.5	0.1	0.1	25		0.1	Pipa et al.

Table 13
HOSPITAL-MANUFACTURED AND COMMERCIALLY AVAILABLE MULTIPLE TRACE-ELEMENT FORMULATIONS

Ion	Shil's			Hull's			Rhode Island Hospital		
	Ion conc. (mg/mℓ)	Salt	Salt conc. (mg/mℓ)	Ion conc. (mg/mℓ)	Salt	Salt conc. (mg/mℓ)	Ion conc. (mg/ℓ)	Salt	Salt conc. (mg/ℓ)
Zinc (Zn)	2.0	$ZnCl_2$	4.13	0.04	$ZnSO_4 \cdot 7H_2O$	0.176	2.0	$ZnCl_2$	4.13
Manganese (Mn)	0.20	$MnSO_4 \cdot H_2O$	1.33	0.04	$MnCl_2 \cdot 4H_2O$	0.144	0.40	$MnSO_4 \cdot H_2O$	1.33
Copper (Cu)	0.40	$CuSO_4 \cdot 5H_2O$	4.0	0.022	$CuSO_4 \cdot 5H_2O$	0.086	1.0	$CuSO_4 \cdot 5H_2O$	4.0
Iodide (I)	0.056	NaI	0.06	0.015	KI	0.019	0.056	NaI	0.07
Fluoride (F)									
Iron (Fe)									
Cobalt (Co)				0.014	$CoCl_2 \cdot 6H_2O$	0.056			
Chromium (Cr)									

Ion	Addamel®[b]			PED-EL®[b]			Travenol®[a]			M.T.E.-4®(LyphoMed)[c]		
	Ion conc. (μg/mℓ)	Salt	Salt conc. (mg/mℓ)	Ion conc. (μg/mℓ)	Salt	Salt conc. (mg/mℓ)	Ion conc. (mg/mℓ)	Salt	Salt conc. (mg/mℓ)	Ion conc. (mg/mℓ)	Salt	Salt conc. (mg/mℓ)
Zinc (Zn)	130.8	$ZnCl_2$	0.27	9.81	$ZnCl_2$	0.02	0.4	$ZnCl_2$	0.834	1.0	$ZnSO_4 \cdot 7H_2O$	4.390
Manganese (Mn)	219.6	$MnCl_2 \cdot 4H_2O$	0.79	13.72	$MnCl_2 \cdot 4H_2O$	0.049	0.2	$MnCl_2 \cdot 4H_2O$	0.720	0.1	$MnSO_4 \cdot H_2O$	0.308
Copper (Cu)	31.75	$CuCl_2 \cdot 2H_2O$	0.085	4.76	$CuCl_2 \cdot 2H_2O$	0.0128	0.2	$CuCl_2 \cdot 2H_2O$	0.536	0.4	$CuSO_4 \cdot 5H_2O$	1.570
Iodide (I)	12.7	KI	0.017	1.26	KI	0.0017	0.024	KI	0.03			
Fluoride (F)	94.95	NaF	0.21	14.24	NaF	0.0315						
Iron (Fe)	279.25	$FeCl_3 \cdot 6H_2O$	1.35	27.92	$FeCl_3 \cdot 6H_2O$	0.135				0.004	$CrCl \cdot 6HO$	0.205
Cobalt (Co)												
Chromium (Cr)							0.4	$CrCl_3 \cdot 6H_2O$	0.002			

[a] Not available in North America.
[b] Pharmacia Canada.
Vitrum Sweden.
[c] Also available as concentrate in which each milliliter has: Zn 5 mg, Cu 1.0 mg, Mn 0.5 mg, Cr 0.01 mg.

Table 14
HOSPITAL AND COMMERCIALLY AVAILABLE SINGLE-ENTITY TRACE-ELEMENT FORMULATIONS

	Toronto General Hospital			International Medication Systems®			Abbott Laboratories®			USV Canada Inc.®		
Ion	Ion conc. (mg/mℓ)	Salt	Salt conc. (mg/mℓ)	Ion conc. (mg/ℓ)	Salt	Salt conc. (mg/mℓ)	Ion conc. (mg/mℓ)	Salt	Salt conc. (mg/mℓ)	Ion conc. (mg/mℓ)	Salt	Salt conc. (mg/mℓ)
Zinc (Zn)	3	$ZnSO_4 \cdot 7H_2O$	13.194	4	$ZnSO_4 \cdot 7H_2O$	17.6	1	$ZnCl_2$	2.085	1	$ZnCl_2$	2.090
Manganese (Mn)	0.7	$MnCl_2 \cdot 4H_2O$	2.523	0.5	$MnSO_4 \cdot H_2O$	1.54	0.1	$MnCl_2 \cdot 4H_2O$	0.360	0.1	$MnCl_2$	0.230
Copper (Cu)	0.2	$CuCl_2 \cdot 2H_2O$	0.5368	1.0	$CuSO_4 \cdot 5H_2O$	3.9	0.4	$CuCl_2 \cdot 2H_2O$	1.073	0.4	$CuCl_2$	0.850
Chromium (Cr)	20 μg/mℓ	$Cr(NO_3) \cdot 9H_2O$	0.1500	10 μg/mℓ	$CrCl_3 \cdot 6H_2O$	0.0512	4 μg/mℓ	$CrCl_3 \cdot 6H_2O$	0.0204	0.004	$CrCl_3$	0.0121
Iodine (I)	120 μg/mℓ	KI	0.157									
Selenium (Se)	120 μg/mℓ	Se (metal)	0.120									

each institution. Generally however, solutions are prepared aseptically by dissolving the various salts in normal saline with a preservative or in plain sterile water for injection. The solutions are then sterilized by either sterile membrane filtration directly into presterilized containers or by terminal steam sterilization. Usually solutions are packaged in 10 to 30 mℓ multidose containers. The shelf life of the finished product varies depending on the type of preparation, method of manufacture, and extent of quality control. These solutions have been found stable for a period of up to 1 year. Table 15 outlines the various formulations used at the Toronto General Hospital. Directions for the preparation of these injectables can be obtained from the author.

F. Packaging and Packaging Components

For practical and economic reasons solutions should be packaged in multidose containers in the presence of an antimicrobial preservative. The container should be of glass Type I, borosilicate-coated to assure maximum stability of the solution in the container. Type I glass has low levels of those substances that may be leached from it and so affect the stability of the enclosed solution. This type of glass is also resistant to chemical reaction with the contents. These containers must be washed with a hot detergent and rinsed thoroughly with pyrogen-free distilled water prior to use.

Rubber closures for sealing the multidose container must be tested for compatibility with the packaged solution. Trace elements such as zinc, iron, and manganese are normally present in a variety of these rubber closures and thus can be leached out into the packaged solution. Furthermore, these closures can adsorb substances from solutions, thus further affecting the concentration of trace elements in the solution.

A recent study at the Toronto General Hospital has shown that the cleaning process and the type of rubber closure have an effect on the presence of particulate matter in the trace element solution. Results of the study showed that the most effective way of cleaning the rubber closures is by boiling them in 3% DET® detergent for 30 min, followed by thorough rinsing with hot distilled water and finally with filtered distilled water using a 0.45-μm membrane filter. Grey butyl rubber stoppers No. 1720 and 1880 (West Co. Ltd., St. Petersburg, Fla.) were the most acceptable. By contrast, natural rubber stoppers were the least acceptable. Large amounts of zinc were leached from the natural rubber closures which were of red and pink colors. Zinc, chromium, copper, manganese, selenium, and iron solutions were found to be stable with grey butyl rubber closures for a period of up to one year. If the manufacturing technique involves the use of commercially available presterilized vials, the type of closure on the vial should be examined for compatibility with the packaged trace element solution.

G. In-Hospital Quality Control of Trace-Element Solution

Where commercial products are used, the need for quality control is minimal. The product monograph should be consulted regarding use and storage of the product. However, where these injectables are manufactured by the institution, it is essential that proper quality control be applied at all stages of the preparation of these solutions. Ingredients used must be of high purity. A qualitative and quantitative analysis record must be made available by the manufacturer. When feasible, the ingredient should be chemically identified and tested for pyrogens prior to use.

Equipment used to manufacture the solutions must be made pyrogen free by either washing with a hot alkaline detergent and rinsing thoroughly with pyrogen-free distilled water or by using dry heat at 250°C for 1 hr.

The final solution must be checked for the presence of particulate matter. This can be done by visual examination of the container with the solution, using an inspection lamp.

Sterility and pyrogenicity must be checked to ensure the absence of microbial contamination. Sterility may be checked by the pharmacy department if facilities are available, or by the microbiology department of the institution.

Table 15
TRACE ELEMENT FORMULATIONS

Trace element	Chemical source	Supplier	Quality of chemical	Quantity of chemical (mg/mℓ)	Trace element ion conc.[a] (mg/mℓ)	Unit size (MDV)[b]
Copper (Cu^{2+})	Cupric chloride ($CuCl_2 \cdot 2H_2O$)	British Drug House	Analytical reagent (Analar)	0.8052	0.3	10 mℓ
Iodine (as Iodide I^-)	Potassium iodide (KI)	British Drug House	BP and USP	0.157	0.12	10 mℓ
Zinc (Zn^{2+})	Zinc sulfate ($ZnSO_4 \cdot 7H_2O$)	J. T. Baker Chemical Company	Analytical reagent	13.194	3.0	10 mℓ
Manganese (Mn^{2+})	Manganous chloride ($MnCl_2 \cdot 4H_2O$)	British Drug House	Analytical reagent (Analar)	2.523	0.7	10 mℓ
Selenium (Se^{4+})	Selenium metal (Se)	Matheson Coleman & Bell	99.9% Se metal	0.120	0.12	10 mℓ
Chromium (Cr^{3+})	Chromic nitrate [$Cr(NO_3)_3 \cdot 9H_2O$]	Fisher Scientific	Certified reagent	0.150	0.02	10 mℓ
Iron (Fe)	Iron-dextran (Imferon)	Fisons	USP	0.02 mℓ of iron-dextran concentrate[c]	0.1	10 mℓ

Note: Manufactured by the Department of Pharmaceutical Services, Toronto General Hospital.

[a] Concentration currently under review.
[b] MDV = multidose vial.
[c] Both ferric and ferrous form; majority in ferric complex-dextran state.

Pyrogen testing can also be performed by the department using the in vitro Limulus Amoebocyte Lysate (LAL) test (Chapter 16) or done by the official USP test in an outside laboratory.

Physical and chemical stability of the solution must be studied during storage to establish the appropriate shelf life of the product. Chemical stability can be done by an outside laboratory if hospital facilities are not available. Ideally, chemical analysis should be done with every batch of solution manufactured, however, under hospital conditions this may not be economically feasible; thus once the product is formulated, manufactured, and tested, adherence to established procedures must be maintained to be sure of the quality of the finished product.

H. Recommendations for the Use and Extemporaneous Manufacture of Trace-Element Solutions

1. Consideration must be given to the presence of trace elements in the commercial intravenous fluids when therapeutic supplementation of these elements is indicated.
2. Trace element supplementation should not be omitted on the basis of their presence in the intravenous fluids since the levels in these solutions are highly variable.
3. Rubber closures must be suitably cleaned prior to being used for the packaging of trace elements.
4. Grey butyl rubber stoppers No. 1720 are a relatively good closure: inexpensive and compatible with the trace metals.
5. Multidose vials to be used for the packaging of the solution must be properly cleaned prior to use.
6. Multidose vials including presterilized ones must be of Type I glass, preferably borosilicate-coated.
7. Until proven unnecessary, antimicrobial preservatives should be included in the trace-element formulations packaged in multidose vial containers.
8. Benzyl alcohol 0.9% is a suitable preservative for trace-element formulations.
9. Solutions of trace elements should be terminally sterilized by steam except in the case of heat-labile products.
10. Adequate quality control should be performed to be certain of the stability and sterility of the solution during the manufacturing process and storage.
11. Thorough admixing must be done following addition of the trace-element solution to a TPN solution.

READING LIST

GENERAL

Bivins, B. A., Rapp, R. P., Record, K., Meng, H. C., and Griffen, W. O., Parenteral safflower oil emulsion (Liposyn 10%) safety and effectiveness in treating or preventing essential fatty acid deficiency in surgical patients, *Ann. Surg.*, 191, 307, 1980.

Brian, L., Intralipid compatibility study: letters, *Drug Intell. Clin. Pharm.*, 8, 75, 1974.

Christensen, H. N., Lynch, E. L., Decker, D. G., and Powers, J. H., The conjugated non-protein, amino acids of plasma. IV. A difference in the utilization of the peptides of hydrolysates of fibrin and casein, *J. Clin. Invest.*, 26, 849, 1947.

Cannon, J. J., Diarrhea possibly caused by total parenteral nutrition (letter), *N. Engl. J. Med.*, 301, 273, 1979.

Deitel, M. and Kaminsky, V., Total nutrition by peripheral vein — the lipid system, *Can. Med. Assoc. J.*, 111, 152, 1974.
DiMatteo, F. P., Dowling, R. J., and Berkowitz, H., Use of standard electrolyte additives in total parenteral nutrition program, *Hosp. Pharm.*, 12, 27, 1977.
Dudrick, S. J., MacFadyen, B. V., Jr., Buren, C. T. Van, Jr., Ruberg, R. L., and Maynard, A. T., Parenteral hyperalimentation: metabolic problems and solutions, *Ann. Surg.*, 176, 259, 1972.
Elsberry, V. A., Grangeia, J. M., Giorgianni, S. J., Jr., and Turco, S. J., The lipid phase in T. P. N., *Am. J. I. V. Ther.*, 23, 1977.
Finkbiner, B. R., Arcos, M., and Ahmed, S. Z., Arterial and venous ammonia response to intravenous protein hydrolysate administration, *Metabolism*, 11, 1077, 1962.
Frank, J. T., Intralipid compatibility study, *Drug Intell. Clin. Pharm.*, 7, 351, 1973.
Geyer, R. P., Parenteral nutrition, *Physiol. Rev.*, 40, 150, 1960.
Ghadimi, H. and Kumar, J., High ammonia content of protein hydrolysates, *Biochem. Med.*, 5, 548, 1971.
Grand, J. P., Cox, C. E., Kleinmay, L. M., Maher, M. M., Pittman, M. A., Tangrea, J. A., Brown, J. H., Gross, E., Beazley, R. M., and Jones, R. S., Serum hepatic enzyme and bilirubin elevations during parenteral nutrition, *Surg. Gynecol. Obstet.*, 145, 573, 1977.
Hakansson, I., Physico-chemical changes in artificial fat emulsions during storage, *Acta Chem. Scand.*, 20, 2267, 1966.
Hallberg, D., Therapy with fat emulsion, *Acta Anaesthesiol. Scand.*, Suppl. 55, 131, 1974.
Heird, W. C., Dell, R. B., Driscoll, J. M., Jr., Grebin, B., and Winters, R. W., Metabolic acidosis resulting from the intravenous alimentation mixtures containing synthetic amino acids, *N. Engl. J. Med.*, 287, 943, 1972.
Johnston, J. D., Albutton, W. L., and Sunshine, P., Hyperammonemia accompanying parenteral nutrition in newborn infants, *J. Pediatr.*, 81, 151, 1972.
Long, C. L., Zikria, B. A., Kinney, J. M., and Geiger, J. W., Comparison of fibrin hydrolysates and crystalline amino acid solutions in parenteral nutrition, *Am. J. Clin. Nutr.*, 27, 163, 1974.
Madan, P. L., Madan, K. D., and Palumbo, J. F., Total parenteral nutrition, *Drug Intell. Clin. Pharm.*, 10, 684, 1976.
Meng, H. C., Use of fat emulsions in parenteral nutrition, *Drug. Intell. Clin. Pharm.*, 6, 321, 1972.
McNiff, L. B., Clinical use of 10% soybean oil emulsion, *Am. J. Hosp. Pharm.*, 34, 1080, 1977.
Pharmacia (Canada) Ltd., Nutralipid 10% (monograph), Dorval, Quebec, 1979.
Phillips, G. D. and Odgers, C. L., *Parenteral and Enteral Nutrition — A Practical Guide*, Natl. Lib. Aust., Flinders University of South Australia, Adelaide, 1980.
Solassol, C., Joyeux, H., Serron, B., Pujol, H., and Romien, C., Nouvelles techniques de nutrition parenterale a long terme pour supplement intestinale, *Chirurgie*, 105, 15, 1973.

TRACE ELEMENTS

Nutritional Advisory Group, Guidelines for essential trace element preparation for parenteral use, *JAMA*, 241, 2502, 1979.
Allisson, R., Plasma trace elements in total parenteral nutrition, *J. Parenteral Enteral Nutr.*, 2, 35, 1978.
Anon., Zinc, *Med. Lett.*, 20, 57, 1978.
Becroft, D. M., Dix, M. R., and Farmer, K., Intramuscular iron-dextran and susceptibility of neonates to bacterial infections, *Arch. Dis. Child.*, 52, 778, 1977.
Bernstein, B. and Layden, T. J., Zinc deficiency and acrodermatitis after intravenous hyperalimentation, *Arch. Dermatol.*, 114, 1070, 1978.
Carlson, J., Kennedy, E., and Melethil, S., Trace element content of Aminosyn hyperalimentation solutions, (Abstr.) poster presentation at 13th Ann. Midyear Clin. Conf. Pharmacists, Las Vegas, 1979.
Capper, K. R., Interaction of rubber with medicaments, *J. Mond. Pharm.*, 4, 305, 1966; quoted in **Boyett, J. B. and Avis, K. E.**, *Bull. Parent. Drug Assoc.*, 29, 1, 1975.
Dowdy, R. P. L., Copper metabolism, *Am. J. Clin. Nutr.*, 22, 887 1969.
Dudrick, S. J. and Rhoads, J. E., New horizons for intravenous feeding, *JAMA*, 215, 939, 1971.
Fliss, D. M. and Lamy, P. P., Trace elements and total parenteral nutrition, *Hosp. Formul. Manage.*, 14, 698, 1979.
Freund, H., Atamian, S., and Fischer, J. E., Chromium deficiency during total parenteral nutrition, *JAMA*, 241, 496, 1979.
Hauer, E. C. and Kaminski, M. V., Trace metal profile of parenteral nutrition solutions, *Am. J. Clin. Nutr.*, 31, 264, 1978.
Heller, R. M., Kirschner, S. G., O'Neill, J. A., Jr., Hough, A. J., Jr, Howard, L., Khommer, S. S., and Green, L. H., Skeletal changes of copper deficiency in infants receiving prolonged total parenteral nutrition, *J. Pediatr.*, 92, 947, 1978.
Hoffman, R. P. and Ashby, D. M., Trace element concentrations in commercially available solutions, *Drug. Intell. Clin. Pharm.*, 10, 74, 1978.

Hull, R. L., Use of trace elements in intravenous hyperalimentation solutions, *Am. J. Hosp. Pharm.*, 31, 759, 1974.
Hull, R. L. and Cassidy D., Trace element deficiencies during total parenteral nutrition, *Drug. Intell. Clin. Pharm.*, 11, 536, 1977.
Jeejeebhoy, K. N., Chu, R. C., Marliss, E. B., Greenberg, G. R., and Bruce-Robertson, A., Chromium deficiency, glucose intolerance, and neuropathy reversed by chromium supplementation in a patient receiving long-term total parenteral nutrition, *A. J. Clin. Nutr.*, 30, 531, 1977.
Jetton, M. M., Sullivan, J. F., and Burch, R. E., Trace element contamination of intravenous solutions, *Arch. Intern. Med.*, 136, 782, 1976.
Kerkhof, K., Trace element concentrations in Travasol injections, letter to the Editor, *Drug. Intell. Clin. Pharm.*, 11, 690, 1977.
Liorda, J. F., Biochemistry of zinc, *Med. Clin. N. Am.*, 60, 661, 1976.
Louria, D. B., Joselou, M. M., and Brawder, A. A., The human toxicity of certain trace elements, *Ann. Intern. Med.*, 76, 307, 1972.
Matol, J. R. and Jeffry, L. P., Formulation of a trace element solution for long-term parenteral nutrition, *Am. J. Hosp. Pharm.*, 35, 165, 1978.
Muller, R. J. and Pipp, T. L., Modern clinical nutrition, the role of trace elements in intravenous nutrition, *Am. J. I.V. Ther. Clin. Nutr.*, p. 14, 1980.
National Dairy Council, The role of essential trace elements, *Dairy Counc. Dig.*, 44, 19, 1973.
Odne, M. A., Lee, S. C., and Jeffry, L. P., Rationale for adding trace elements to total parenteral nutrient solutions — a brief review, *Am. J. Hosp. Pharm.*, 35, 1057, 1978.
Orten, J. M. and Henhaus, O. W., Eds., *Biochemistry*, 8th ed., C. V. Mosby, St. Louis, 1970, 437.
Pharmacia (Canada) Ltd., PED-EL package insert, Dorval, Quebec, 1979.
Pharmacia (Canada) Ltd., ADDAMEL package insert, Dorval, Quebec, 1979.
Pipa, D. A. and Tsallas, G., Trace Elements — Pharmaceutical Considerations, Hospital Pharmacy Residency Project, Toronto General Hospital, Canada, 1980.
Reinhold, J. G., Trace elements — a selective survey, *Clin. Chem. (Winston-Salem, N.C.)*, 21, 476, 1975.
Roulet, M., Relationship between iron status and incidence of infection in infancy, *Pediatrics*, 62, 246, 1978.
Sandstead, H. H., Burk, R. F., Booth, G. H., and Darby, W. J., Current concepts on trace minerals, *Med. Clin. N. Am.*, 54, 1509, 1970.
Shearer, C. A. and Bozian, R. C., The availability of trace elements in intravenous hyperalimentation solutions, *Drug Intell. Clin. Pharm.*, 11, 465, 1977.
Shike, M. and Jeejeebhoy, K. N., Trace Elements and Vitamins in Total Parenteral Nutrition, Toronto General Hospital, Canada, 1980.
Shike, M., Roulet, M., Kurian, R., Whitewell, J., Stewart, S., and Jeejeebhoy, K. N., Copper requirements in total parenteral nutrition, *Gastroenterology*, 81, 290, 1981.
Shils, M., Wright, W., and Turnbull, A., Long-term parenteral nutrition through an external arteriovenous shunt, *N. Engl. J. Med.*, 283, 341, 1970.
Shils, M. E., Minerals in total parenteral nutrition, *Drug Intell. Clin. Pharm.*, 6, 385, 1972.
Shils, M. E., More on trace elements in total parenteral nutrition solutions, *Am. J. Hosp. Pharm.*, 32, 141, 1975.
Smith, J. L. and Goos, S. M., Selenium nutriture in total parenteral nutrition: intake levels, *J. Parenteral Enteral Nutr.*, 4, 23, 1980.
Solomons, N. W., Layden, T. J., Rosenberg, I. H., Vokhacstu, K., and Sandstead, H. H., Plasma trace metals during total parenteral nutrition, *Gastroenterology*, 70, 1022, 1976.
Ulmer, D. D., Trace elements, *N. Engl. J. Med.*, 297, 318, 1977.
Underwood, E. J., *Trace Elements in Human and Animal Nutrition*, 4th ed., Academic Press, New York, 1977.
Utter, M. F., The biochemistry of manganese, *Med. Clin. N. Am.*, 60, 713, 1976.
van Caillie, M., Luijendijk, I., Degenhart, H., and Fernandes, J., Zinc content of intravenous solutions, *Lancet*, 2, 200, 1978.
van Rij, A. M., Thomson, C. D., and McKenzie, J. M., Selenium deficiency in total parenteral nutrition, *Am. J. Clin. Nutr.*, 32, 2076, 1979.
Wan, K. K. and Tsallas, G., Dilute iron dextran formulation for addition to parenteral nutrient solutions, *Am. J. Hosp. Pharm*, 37, 206, 1980.
Wolman, S. L., Anderson, G. H., Marliss, E. B., and Jeejeebhoy, K. N., Zinc in total parenteral nutrition: requirements and metabolic effects, *Gastroenterology*, 76, 458, 1979.

Chapter 14

MANUFACTURE OF TPN SOLUTIONS

George Tsallas

TABLE OF CONTENTS

I. GENERAL CONSIDERATIONS

The method of preparing TPN solutions will vary from hospital to hospital depending on various factors such as facilities available, types of patients, scope of the program, etc. The manufacture of TPN solutions is most safely handled by the Pharmacy department. In evolving such a program, the Pharmacy should consider the following facets of the task:

1. Space and other facilities
2. Inspection and storage of raw materials
3. Equipment
4. Personnel
5. Compounding technique
6. Records
7. Labeling
8. Storage of finished product
9. Quality control
10. Delivery of finished product to the patient care area

II. SPACE REQUIREMENTS AND CONSTRUCTION OF THE COMPOUNDING AREA

Space has to be sufficient to compound, store, and package TPN solutions. Special consideration should be given to the organization and construction of the space used for compounding. This area should be away from heavy traffic, be dust-free, and be amenable to daily cleansing. To meet these objectives it is necessary to have stainless steel counters, shelves, and sinks for hand washing. Areas for the cleaning and storage of large pieces of equipment should also be available. The walls, ceiling, and floors should be made of nonporous material capable of being washed on a regular basis with detergent. In addition to these physical facilities it is necessary to have good lighting and ventilation with dust-free filtered air. An adjacent room should be available to enable personnel to change into sterile gowns. Finally, in this area no paper boxes or dusty objects should be allowed to accumulate.

III. INSPECTION AND STORAGE OF RAW MATERIALS

The containers of commercial stock solutions of amino acids, dextrose, electrolytes, etc., can become damaged during transportation. They must be carefully inspected for any defects such as wet or deformed cartons. Any damaged cartons, as well as those with defaced labels, should be returned to the manufacturer for credit or replacement. Light-sensitive solutions must be retained in their original protective packaging.

Stock solutions must be stored in a designated area and the inventory rotated. The area should be well lighted so as to be able to examine solutions and containers, and the temperature should be controlled to avoid any decomposition.

IV. EQUIPMENT

Equipment needs will vary depending primarily on the method used by the institution to prepare TPN solutions. The following is a list of essential equipment.

- Laminar air-flow hood
- Refrigerator

- Shelving
- Waste containers
- Delivery baskets or carts
- Visual inspection lamp
- A pH meter
- Disposable membrane filters, syringes, and transfer tubing
- Glassware and/or stainless-steel graduated cylinders
- Containers for storage

Other equipment that may be needed for special methods of manufacture and quality control could include a suitable mixing vessel made from either stainless steel or glass, and apparatus using a membrane filter for clarification and sterilization of solutions. In addition a microscope for the detection of particulate contaminants, a hot plate, a stove, and a hot-water bath for pyrogen testing may be required.

V. PERSONNEL

The preparation of TPN solutions demands the expertise of specially trained professional and nonprofessional personnel. The professional pharmacist, specialized in this area of pharmaceutical service, must have a thorough knowledge of aseptic technique and manufacturing procedures. He/she must have access to the necessary equipment for the preparation and quality control of these solutions and the associated devices used for the delivery of TPN. The pharmacist should train both professional and nonprofessional personnel involved with the preparation of TPN solutions. Technical personnel must be trained to follow a rigid protocol and they should be regularly checked and supervised. Professional personnel must upgrade their skills and knowledge in this area by attending educational seminars and specialized courses.

Both professional and technical personnel have distinct roles, responsibilities, and duties. Those of the pharmacist have been discussed in Chapter 1. Those of the pharmacy technician are assigned and monitored by the supervising pharmacist. Such duties include:

1. Cleaning counter tops, transport carts, and preparing the laminar air-flow hood prior to and after use.
2. Cleaning of equipment prior to and after preparation of the solutions.
3. Assembling together all required supplies and transporting them to the compounding area.
4. Preparing labels and labeling of finished product.
5. Checking all ingredients and supplies prior to use by the pharmacist.
6. Compounding solutions under supervision of the pharmacist.
7. Sealing, packaging, and storing solutions.
8. Restocking shelves with supplies.
9. Procuring and distributing supplies to ambulatory patients.
10. Performing other duties as assigned by the pharmacist.

VI. COMPOUNDING TECHNIQUE

It is important to remember that all components used in admixing TPN solutions should be handled with care to maintain the standards set by the manufacturer's quality control procedures. Where the procedure compromises or alters the manufacturer's seal, as in bulk compounding of these solutions, the procedures, and control measures used must ensure sterility and absence of pyrogenicity.

Touch contamination is one of the commonest ways of bacterially contaminating solutions. Any area or item that is considered sterile, or comes in contact with the solution, must not be touched or handled. Hands must be washed thoroughly with a scrub disinfectant as often as necessary. Gloves, gowns, masks, and haircaps should be worn to reduce contamination due to shedding of dry skin, hair, particulate matter, and bacteria. It is critical that knowledge of the proper use of the laminar air-flow hood be put into practice and that interruptions be avoided during compounding. Unnecessary traffic and unauthorized personnel must be restricted from entering the compounding and sterile filling rooms.

VII. RECORDS

Records must be established, maintained, and available pertaining to the manufacture and quality control of all TPN solutions dispensed. The records should contain details about ingredients used, quantities of such ingredients, concentration of all ingredients in the final admixture, volume and number of units prepared, signatures of those preparing and checking the solutions, batch or identification number, preparation date, and type of solution. Quality control records concerning raw material, equipment, and finished product are essential to ensure excellence in patient care. All records are official records and must be easily accessible, kept up-to-date, and retained for at least 5 years.

VIII. LABELING

Labeling of the finished product should include:

1. Patient's name
2. Patient's room number
3. Basic ingredients of the solution
4. Additives and their concentration
5. Batch number or preparation date
6. Total volume
7. Expiry date
8. Auxilliary labels (for example, instructions to refrigerate)

Where additives or other medicines are added by the nurse on the floor, a medication label must be attached to the bottle to identify the additive, dosage, person adding additive, date, and patient's name.

IX. STORAGE OF FINISHED PRODUCT

TPN solutions support bacterial and fungal growth. It is imperative that these solutions be kept refrigerated at 4°C until used, or up to the expiry date. Elevated temperatures increase the chances of bacterial contamination and also potentiate the Maillard reaction between amino acids and dextrose resulting in the development of a dark color. Solutions left out of the refrigerator for longer than 12 hr should not be administered unless checked and authorized by the pharmacy. No other biological material including food should be stored along with these solutions since this practice will increase the chance of bacterial contamination.

X. QUALITY CONTROL

This is an area of great importance in assuring the quality and the safety of the product for human use. It is necessary that the quality of the raw materials be monitored to ensure that the final admixture is safe to administer. Containers of commercial stock solutions must be checked for proper labeling, expiry date, defects in the container, and unusual color change in the solutions. The final admixture must be tested for sterility, absence of particulate matter, and pyrogenicity. Pyrogen testing may not always be feasible, however where bulk compounding procedures are used it is essential that solutions be checked for pyrogens. Visual examination of the final product must be done to ensure the absence of any discernible particulate matter. Furthermore, all equipment must be maintained in good working order and kept clean in order to produce an acceptable final product.

Such equipment includes the laminar air-flow hood (LAFH). Testing of this should include:

1. Measurement of air velocity
2. Determination of air-flow patterns
3. The dioctylphthalate smoke test (DOP)
4. Microbiological checks

The LAFH should be inspected and certified regularly to ensure proper and safe operation. The inspection should be carried out by a reputable firm specializing in such procedures.

It is imperative that prefilters be changed once every month and the HEPA filter tested twice* a year. A microbiological test, which consists of placing blood-agar plates inside the hood both while in use and not in use, is done to show that the system effectively excludes air-borne bacterial contaminants. This procedure should be done at least every two weeks.

In addition, other equipment such as mixing tanks and filtration units used during the manufacture of solutions should be kept clean and in good operating order. Furthermore, such equipment should be restricted solely to the manufacture of sterile products.

The environment also must be monitored by procedures which include microbiological testing of the area where solutions are prepared using culture plates exposed to the atmosphere and the plating of swabs of the working surfaces. The results of these cultures are useful indicators of the degree of microbial contamination of the environment. In addition regular schedules for cleaning ceilings, walls, floors, benches, etc. as well as for changes of air filters must be established and maintained.

The adherence of personnel to proper procedures and controls in all areas of solution manufacture should be monitored, and those not following procedural guidelines should not be allowed to continue to perform their duties. Furthermore, continuing-education programs should be implemented and attended by the personnel for updating purposes.

All quality control measures must be documented and the records kept for official purposes.

Quality control must be exercised daily, in all areas of solution manufacture.

XI. DELIVERY OF FINISHED PRODUCT TO THE PATIENT CARE AREA

The personnel responsible for the transportation of TPN solutions to the nursing unit must be properly instructed as to the handling and storage of these solutions. Any

* Once a year has been found to be satisfactory for certain institutions that do not operate the LAFH 24 hr/day.

units damaged during transportation must be brought back to the pharmacy and discarded. Any solutions returned to the pharmacy should be packaged prior to transport from the nursing units. All returned TPN solutions must be checked by a pharmacist in regard to storage conditions prior to return, physical state, label, and expiry date. Solutions that are found to have been properly stored and labeled and have not been physically damaged can be restocked and be available for use until the expiry date. Solutions left unrefrigerated for an unknown period of time or those having passed the expiry date must be discarded.

READING LIST

Burke, W. A., Preparation, including incompatibilities and instability, in *Symposium on Total Parenteral Nutrition,* American Medical Association, Chicago, 1972, 175.

Edwards, D. D. and O'Brien, T. E., Preparing single unit hyperalimentation solutions in a small hospital, *Hosp. Pharm.*, 7, 46, 1972.

Giovanoni, R., *The Manufacturing Pharmacy Solutions and Incompatibilities in Total Parenteral Nutrition,* Fisher, J. E., Ed., Little, Brown, Boston, 1976.

Grant, J. N., Moir, E., and Fago, M., Parenteral hyperalimentation, *Am. J. Nursing,* 69, 2392, 1969.

Klotz, R., Sherman, J. O., and Egan, T., Preparation of hyperalimentation solutions for the pediatric patient, *Am. J. Hosp. Pharm.*, 28, 102, 1971.

National Coordinating Committee on Large Volume Parenterals, Recommended methods for compounding intravenous admixtures in hospitals, *Am. J. Hosp. Pharm.*, 32, 261, 1975.

Pike, M., Whitaker, H., and Grossman, A., A simplified method of preparing hyperalimentation fluids, *Pharm. Times,* 35, 48, 1969.

Sasamoto, R. M., Hollenbeck, D. E., and Melnick, N., Preparation of sterile parenteral hyperalimentation solutions by membrane filtration, *Am. J. Hosp. Pharm.*, 28, 370, 1971.

Stolar, M. H., Assuring the quality of intravenous admixture programs, *Am. J. Hosp. Pharm.*, 36, 605, 1979.

Tsallas, G. and Baun, D. C., Home care total parenteral alimentation, *Am. J. Hosp. Pharm.*, 29, 840, 1972.

Chapter 15

PROCEDURES AND COMPOUNDING TECHNIQUES

George Tsallas

TABLE OF CONTENTS

I. INTRODUCTION

Basically there are two methods presently used by the majority of hospitals engaged in the preparation of TPN solutions. These are individually prepared solutions (vacuum bottle technique and plastic bag technique), and bulk compounded solutions.

II. INDIVIDUALLY PREPARED SOLUTIONS

A. Vacuum Bottle Technique

This method of compounding solutions (Figure 1) may depend on either commercially available hyperalimentation or TPN "kits" or the use of empty sterile evacuated bottles of various sizes.

1. Kit Method

The system using TPN kits provided by drug firms comprises (a) a 1-ℓ bottle partly filled with 500 mℓ of 50% sterile dextrose and sufficiently evacuated so as to allow the transfer to it of other solutions, (b) a bottle of an amino acid solution of variable concentration depending on the manufacturer, and (c) a transfer set and a sterile protective cap. The equipment referred to above comes with detailed diagrams and step by step instructions for performing the admixing.

2. Sterile Bottle Technique

The system using empty sterile evacuated bottles consists of (a) a commercially available sterile evacuated container of the desired volume, (b) solutions of concentrates such as dextrose 50% or amino acids, and (c) transfer sets and protective caps.

3. Technique of Admixture

The technique of admixing in both of the above systems is similar. Usually the two basic components, amino acids and dextrose, are first admixed in the laminar air-flow hood (LAFH). The admixing process is accomplished using a transfer set. One end of

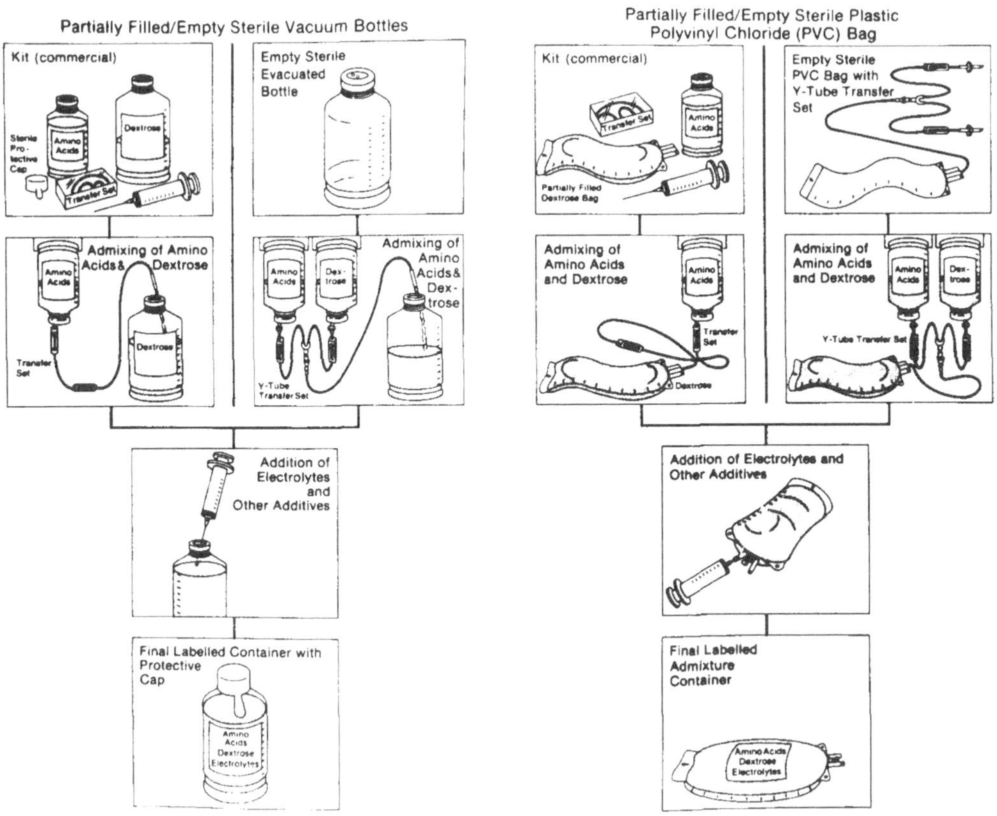

FIGURE 1. Flow diagram to show steps in the various individual methods of preparation of parenteral nutrition solutions.

the transfer set with the flow rate clamp closed is aseptically inserted into the appropriate entry port site of a full bottle containing either the amino acids or the dextrose. The other end of the transfer set is aseptically inserted through the appropriate entry port of either the partially filled container or the empty sterile evacuated container. The full unit is next inverted and suspended and the flow clamp on the transfer set is opened to allow transfer of solution by gravity into the partially filled container or empty sterile evacuated unit. When transfer of solutions is completed the flow clamp is closed and the transfer set is aseptically removed from the container with the final admixture of the two basic components, leaving the large-bore needle of the set in place in the stopper of this container.

Electrolytes and other additives are then aseptically added through the large-bore needle using a sterile disposable plastic syringe and needle.

When all additions are completed the needle is removed from the stopper of the final container, the site is "prepped", and the sterile additive cap affixed. A modified method can be applied if the sterile vacuum is to be maintained. In this case some or all additives can be added aseptically to the full amino acid bottle through the rubber diaphragm prior to admixing with the dextrose in the partially filled bottle or transferred into the empty evacuated bottle.

The final admixture is examined for particulate contamination, labeled, and refrigerated until use. Since the admixture has not been finally sterilized it is advisable to use it within 24 hr of its preparation, particularly if it contains vitamins or other drug additives such as insulin, that are labile in nature. Although, physicochemically, solutions prepared in this manner are found to be stable for a period longer than 24 hr, it is advisable to use such solutions within 24 hr. The final admixture container is not a completely closed unit and thus may be relatively easily invaded by microorganisms. Thus for greater safety in avoiding possible microbial contamination such solutions should not be stored longer than 24 hr.

B. Plastic Bag Technique

This technique is similar to the vacuum bottle technique except that either empty sterile plastic bags of polyvinyl chloride (PVC) or partially filled bags are used.

1. Method

As in the vacuum bottle technique there are commercial "TPN kits" basically consisting of partially filled 1-ℓ plastic bags containing either 500 mℓ or 750 mℓ of dextrose 50% or 33.3%, a full bottle of an amino acid solution, and a transfer set.

The empty sterile PVC bags are either fitted with a single filling tube or with a Y-tube, spike (needles), connectors, flow-rate adjustment clamps, and an injection port for the addition of additives. Usually the admixing technique in both cases only involves the admixing of the two basic components, amino acids and dextrose, by aseptic transference of one component either into a partially filled unit containing the other component or into a sterile empty PVC unit. If the empty sterile PVC bags has a Y-tube filling system, both components (amino acids and dextrose) may be simultaneously transferred and admixed together in the empty bag. With this technique a Viavac® unit (Baxter Laboratories) can be used to produce a vacuum in the partially filled or empty bag and thus speed the transfer of the solutions.

All required additives are added aseptically to the final admixture and mixed thoroughly. As with the vacuum bottle method, a modified procedure may be applied where the additives can be added to one component, usually the amino acids, then admixed with the dextrose solution.

The final container is examined for visible particulate matter and squeezed for possible leakage. It is next properly labeled, usually with a 24-hr expiry date. This expiry date is basically a precaution because of possible microbial contamination. This system provides a closed environment for the final admixture without need for an air bubble. Solutions containing amino acids, dextrose, and minerals have been found stable with no significant physicochemical changes or microbial contamination for a period of up to 30 days.

Following completion of amino acid-dextrose admixing in the final container, the filling tubing of the container or the transfer set in the case of the commercial kits, is securely sealed near the container with a metal crimp or other acceptable method and the excess tubing is removed.

2. Common Considerations

Both the vacuum bottle and plastic bag techniques have certain things in common. They both utilize partially filled containers or complete empty sterile containers. Both can utilize gravity or vacuum to effect solution transfer and both are available in commercial kit form which can simplify matters. Furthermore, both require similar manipulative steps during the admixing and transferring process.

Various modifications of these methods have been adopted. Some of these include a combination of a vacuum bottle and empty sterile bag technique where the final admixture is transferred from the vacuum bottle container into an empty sterile PVC bag; the use of special injection sites with extension tubing for multiadditive injection to preserve the injection port of the final admixture container and the use of in-line membrane filters (0.45 to 5 μm) to filter out particulate-matter contaminants during the transferring of the amino acid solution with the various additives. This process, although desirable, is slow and time consuming. More recently a specially designed commercial pump, IV-6500 Formulator®, is used to simultaneously admix all three basic components of dextrose, amino acids, and water into a final sterile PVC container. Although investigators have shown this pump to be cost-effective and convenient, additives still have to be introduced through the injection ports of the final container as in the previous methods.

While the vacuum bottle and plastic bag techniques are the most popular and widely used procedures for the preparation of TPN solutions, they suffer from certain disadvantages in that they may be more expensive and time consuming than the bulk method. Their simplicity makes them attractive for most hospitals just starting a TPN program or having only a small one. However, where expansion of the program is requested and where there are personnel and budgetary restrictions, the use of these methods becomes impractical and expensive.

The following account illustrates the use of the bottle and plastic bag techniques for manufacturing individually tailored solutions at the Toronto General Hospital (TGH). The method has been used and tested in this institution for the past ten years and has been found suitable for making such formulations.

C. TGH Methods for Individually Manufactured TPN Solutions

1. Preparation of a Master Formula

1. Transcribe formulation from TPN order sheet (Chapter 9, Table 1) to master formula sheet (Figure 2). This procedure must be done by the pharmacist.
2. Have another pharmacist check the master formula for corrections and completeness.
3. Date and assign a batch number to the formulation and enter it in the records book.
4. Follow outline for compounding of solutions.

2. Assembly and Preparation of Required Equipment

Prepare required equipment such as the LAFH, stainless steel carts, etc. as described subsequently (Section III) for the bulk compounding method. Assemble all required material within reach of the LAFH.

3. Compounding of Solutions

a. Assembly and Preparation of Commercial Stock Solutions

1. Scrub hands thoroughly with soap prior to handling of stock solutions.
2. Remove stock solution bottles and sterile water-for-irrigation bottles from the cartons and check for defects such as cracks, unusual color, improper labeling, etc.; then place on a clean stainless-steel cart.
3. Take stock solutions as well as the sterile water-for-irrigation bottles into the compounding area and rinse thoroughly with distilled water.
4. Towel-dry all bottles; then clean the tops with ethyl alcohol 95%. (It is appreciated that 70% ethyl alcohol is slightly more germicidal. For convenience however, TGH has used 95% instead and have found it entirely satisfactory for over a decade for the purposes outlined throughout this text.)
5. Have a pharmacist check all stock solution bottles and additives prior to opening.

b. Compounding

1. Attach the master formula to the side of the LAFH so that it will be visible and easy to follow during compounding.

Name of Patient_______________

Solutions: Amino Acids_____% Dextrose_____%

Ward________ Physician__________

TORONTO GENERAL HOSPITAL - PHARMACY DEPT.

TPN MASTER FORMULA WORKSHEET

Pharmacy Batch No.________ Date Made________

Formula Written By_______________

Formula Checked By_______________

	Solution and Additive	Manufacturer	Lot No.	Volume ml or l	No. of Units (Est.)	No. of Units (Act.)	Meas'd & Added By	Checked By	Cost $
Solutions	1.								
	2.								
	3.								
Additives	1.								
	2.								
	3.								
	4.								
	5.								
	6.								
	7.								
	8.								
	9.								
	10.								
	11.								
	12.								

SAMPLE LABEL

(Attach sample label on back)

Label prepared by__________

Label checked by__________ (Pharmacist)

ELECTROLYTE CONTENT (mM)

Ion	Per Unit Sol'n	Per Unit Add.	Total Per Unit	Total Per 24hr[a]
Na				
K				
Ca				
Mg				
Cl				
Acetate				
Lactate				
PO_4				
Other				
CHO Calories				

Accessories	Name	Manufacturer	Lot No.	No. Used	Cost $
Final Container					
Compounding Container					
Transfer Set					
Injection Sites					
Filters					
Other					

Vol. (ml)...... Wt. (g)......of Unit.

No. of Units Prepared...............

TESTS PERFORMED

Sterility.... Pyrogenicity.... Biochemistry....

Other........................

SOL'N UNITS ADMIN'D PER DAY..................

Date Released__________ Time: Pharmacist__________Hrs.

Supervisor__________ Technician__________Hrs.

Revised: February 1982

a) May include additives in addition to those added by pharmacy.

FIGURE 2. Master formula form used by the Pharmaceutical Services of the Toronto General Hospital for individually prepared TPN formulations.

2. Scrub hands with a disinfectant detergent (e.g., chlorhexidine gluconate 2% skin cleanser).
3. Put on a gown, hair cap, and mask.
4. Take stock solutions, admixing vessels, additives, and a clean 4-ℓ stainless-steel pot inside the sterile filling room.
5. Clean the tops of all bottles with 95% ethyl alcohol.
6. Place the clean 4-ℓ stainless-steel pot inside the LAFH.
7. Take the sterile water-for-irrigation bottles inside the LAFH; then remove the screw caps from each bottle one at a time and ensure that the previously capped area is not touched. Hold the screw cap in one hand, making sure that nothing comes in contact with the inside lining.
8. Pour the sterile water, again without touching the mouth of bottle and also without splashing, into the clean empty vessel (Figure 3), then discard it leaving an empty sterile bottle.
9. Replace the screw cap on the bottle again without touching the mouth of the bottle. In a similar manner (Steps 7 and 8), empty the sterile water from the other bottles. Do not remove the empty sterile water-for-irrigation bottles to outside the LAFH.
10. Take the amino acid bottle of the stock solution inside the LAFH and remove the metal tops.
11. Scrub the rubber diaphragm of the amino acid bottle with 95% ethanol and let dry before removing diaphragm. Remove rubber diaphragm.
12. Take a sampling injection site (Figure 4), remove it from its protective cover, and insert it aseptically through the large opening of the amino acid bottle.
13. Remove aseptically the rubber diaphragm of the injection site by pulling it off with one hand while supporting the bottle with the other.
14. Aseptically pour from the amino acid bottle into the empty sterile bottle (which has been previously emptied in Step 8) a measured volume of amino acid solution using the graduated scale on the empty bottle.
15. Replace the screw cap on the receiving bottle and proceed as in Step 11 with the other bottles.
16. After adding the amino acid solution to all the bottles, proceed in a similar manner (Step 11) with the dextrose solution, using the appropriate transfer set (e.g., Bot-O-Jet® [IMS] or Decanting Set® [Abbott]).
17. Take electrolyte solutions (either manufactured in the pharmacy or commercially bought) and wipe caps with 95% ethanol. Then remove caps from bottles, ensuring all manipulations remain inside the LAFH, and attach a presterilized secondary irrigation set top (Abbott) (Figure 5) with an injection site.
18. Hang the electrolyte bottle, then using an appropriate sterile disposable syringe, withdraw the required volume of electrolyte solution and add this to the amino acid-water-dextrose mixture. Withdraw additives from ampoules using a monoject needle with a filter to remove possible glass particles. In the case where multidose vials are used instead of ampoules as a source of additives, the following procedure should be employed. Scrub rubber tops of the multidose vials of additives with alcohol and pierce the stopper with a needle at a 45° angle, with the bevel of needle facing upward in order to avoid coring. Mix well by swirling the bottles after the injection of each additive.
19. After all necessary electrolytes and other additives have been injected bring the mixture to volume with sterile water-for-injection applying Steps 8, 9, 10, and 12. Mix well by swirling the bottle.

FIGURE 3. Pouring water-for-irrigation from 1000-m*l* screw-cap bottle into steel 4-*l* mixing pot in the LAFH.

c. Packaging of the Finished Mixture

1. Remove screw caps from the admixture bottles and replace with secondary irrigation set tops (Abbott) (Figure 5).
2. Remove the 1-*l* TA-10 transfer packs (Fenwal®) from their boxes and assemble them in the LAFH.
3. Remove the protective sheath from the spike at the end of the transfer pack tubing and insert it in the opening of the secondary irrigation set attached to the admixture bottle.
4. Hang bottle and allow the solution to empty by gravity into the transfer pack (Figure 6).
5. When the admixture bottle is empty carefully squeeze (Figure 7) the full transfer pack to remove residual air, then clamp the secondary irrigation-set tubing and crimp transfer-pack tubing with a hand sealer clip (Fenwal®) (Figure 8).
6. To fill other bags aseptically remove the spike of the transfer-pack tubing from the secondary irrigation set and attach the spike of another empty transfer pack.
7. Remove the secondary set from the empty bottle and attach to another full amino acid-dextrose-electrolyte bottle and proceed as in Steps 4, 5, and 6.
8. Collect a small sample of solution in a presterilized bottle and send for sterility check, osmolality, and pH determinations. (Ideally, collect a sample from every bottle.)

FIGURE 4. Sterile sampling injection site (insertable spout) thrust into insertion site of 500 m*l* bottle (see Chapter 11, Figure 5) and used to pour its contents into a partly filled, open-mouthed 1000 m*l* bottle in the LAFH.

d. Labeling of Finished Product

1. Prepare labels containing batch number, ingredients, patient's name, nursing unit, volume, expiry date, and auxiliary label (with instruction to refrigerate).
2. Transport filled units of TPN solutions into the prepackaging area.
3. Trim off excess tubing from full transfer pack using clean disinfected scissors.
4. Place filled unit on a clean stainless-steel counter and label.
5 Place labeled units in clean plastic bags and seal with the thermosealer.
6. Package full, finished units into cartons.

e. Storage

1. Store all packaged bags and bottles in the walk-in refrigerator and enter on record sheet.
2. Store boxes with solutions on shelves and not on the floor.
3. Have label on box facing outward for ease of identification.

f. Transport to the Nursing Unit

Follow same procedure as outlined subsequently in this chapter in the section on transport to the nursing unit for bulk compounded TPN solutions (Section III, B.5.f).

FIGURE 5. After the 1000 ml bottle is filled in the LAFH a secondary irrigation set (including an injection site) is attached to allow its subsequent administration.

III. BULK-COMPOUNDED SOLUTIONS

When certain formulas are standardized for use within the hospital or are not often altered as in the case of home care TPN formulations, the most practical and economical means of preparing solutions is on a bulk basis.

Basically this method involves the aseptic admixing of sterile solutions of amino acids, dextrose, and minerals in a pyrogen-free stainless-steel or glass mixing tank. The admixture is then sterilized by pumping it through a series of membrane prefilters and finally through a sterile 0.22 μm membrane into sterile containers. These sterile containers are either sterile glass bottles or sterile empty plastic PVC bags.

Both sterility and pyrogen testing are carried out with each batch. The number of samples tested for sterility and pyrogens will depend on the batch size. The final admixture is also analyzed for electrolyte content and, where feasibile, amino acid and dextrose content.

Additives, other than electrolytes and trace elements, are injected aseptically in a LAFH immediately prior to administration of the TPN solution.

While solutions can be refrigerated for a period of up to 60 days, this span will be affected by the type of container and additive. Basic solutions packaged in PVC bags

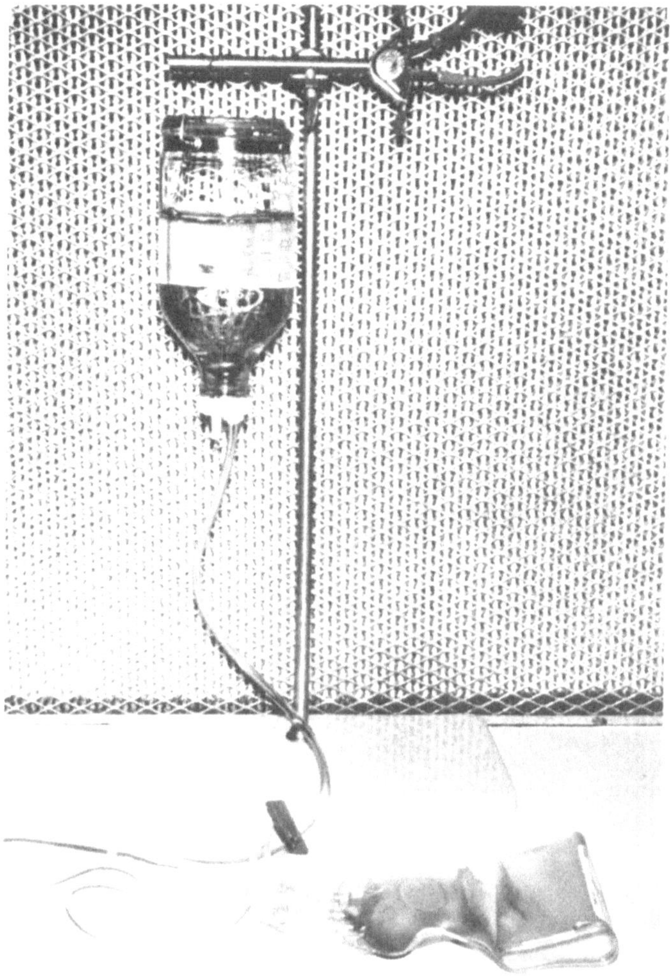

FIGURE 6. Filled bottle of nutrient mixture after mixing now empties by gravity via transfer set into empty transfer pack (flexible polyvinylchloride bag) in the LAFH.

containing amino acids, dextrose, and minerals can be stored with refrigeration for up to 60 days, whereas solutions with more labile additives such as vitamins or insulin should not be stored longer than 24 hr.

This method has many advantages, particularly in accommodating a large and expanding TPN program. Some of these advantages include:

1. Savings in cost and time.
2. Clarification and sterilization of the final admixture.
3. Flexibility for the physician in altering electrolytes and for the pharmacist in preparing these solutions.
4. Minimal waste when changes occur in the electrolyte needs of the patient.
5. Reduction of the manipulative steps and multiple entries to the final unit that are required with the individual bottle method.
6. A longer storage expiry date for final solutions.

The following is a detailed outline of the steps involved with the bulk compounding technique currently used at TGH, beginning with a list of basic requirements (see Figure 9).

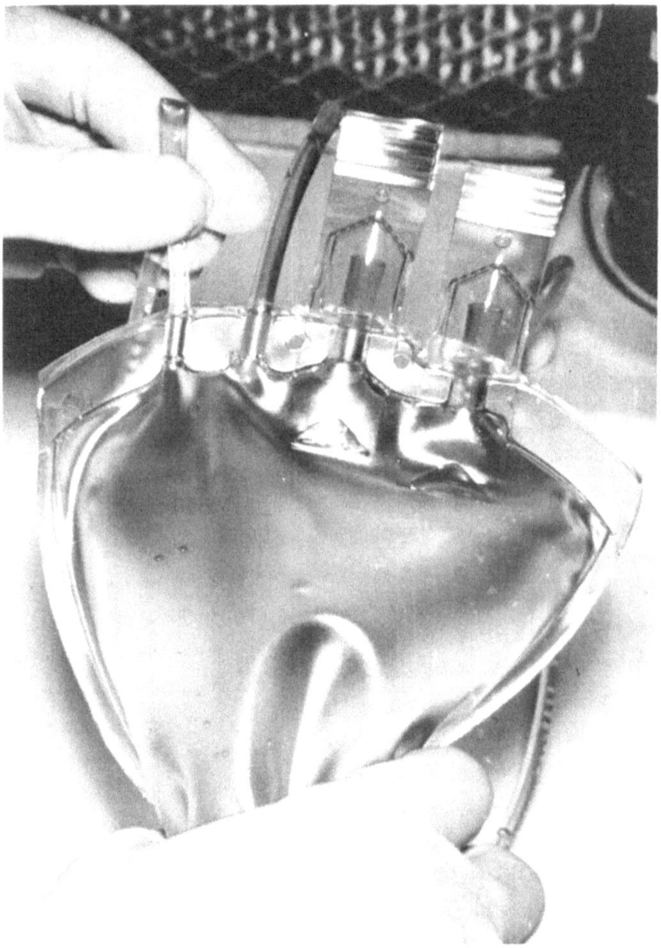

FIGURE 7. Squeezing full transfer pack to remove air prior to sealing. Such removal reduces oxidative changes in the nutrient mixture.

A. Physical Requirements for Bulk Method

1. Equipment

Stainless-steel carts (ensure enough are available)
Stainless-steel tank for admixing
Peristaltic pump
Pressure gauge with manifold
Nitrogen gas
Tygon®/PVC/silicone tubing
Electric stirrer
Stove or hot plate(s)
Steel/glass graduated cylinders
Laminar air-flow hood (LAFH)
Top-loading balance
142 mm/293 mm Filter holder
142 mm/293 mm Membrane filters (various pore sizes)
0.22 μm Sterile filter unit
Steel pliers or bottle decappers

FIGURE 8. Kinking of the tubing attached to filling port of full transfer pack after expelling air as in Figure 7, just prior to maintaining kink or seal (i.e., sealing tubing with an encircling metal crimp using a hand or air crimper).

Hemostat(s)
Plastic cables and tie gun
Stainless-steel scissors
Hand sealer and metal clip
Plastic bag heat sealer
Particle matter examination station
Walk-in refrigerator
Empty PVC bags with extension tubing
Y-shaped glass tubing connectors

2. Solutions

Commercially available concentrated solutions of amino acids, dextrose, minerals, and water.

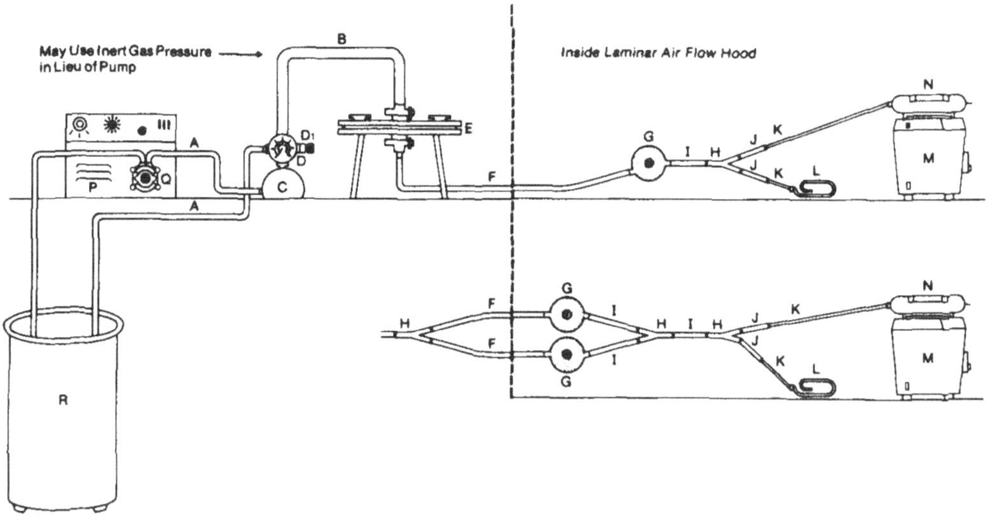

FIGURE 9. Schematic sketch of main components of bulk compounding method used by the Pharmaceutical Services at the Toronto General Hospital, with some alternative steps.

A. Silicone ¼" medical grade tubing
B. Silicone ½" medical grade tubing
C. 5-way manifold
D. Pressure gauge
D1. Bypass valve
E. 142-mm Filter holder with 1.2 μm and 3 μm membrane filters
F. Tygon ¼" tubing
G. Twin-90 (0.22 μm sterile unit)
H. Y-glass connector
I. Sterile Tygon ¼" tubing
J. Sterile Tygon 3/16" tubing
K. TA-10 transfer pack extension tubing
L. Empty TA-10 transfer pack
M. Mettler top-loading balance
N. Full TA-10 transfer pack
P. Variable-speed pump
Q. Pump head
R. Stainless-steel mixing tank

3. Space and Facilities

Admixing area
Sterile filling area
Packaging and labeling area
Refrigerated storage area
Steel counter top
Equipment washing area
Handwashing area
Personnel gowning area

B. Bulk Manufacture of Standard and Home-Care TPN Solutions at TGH

1. Preparation of Master Formula

1. Obtain and/or fill out a master formula sheet (Figure 2). This procedure must be done by the pharmacist.
2. Have another pharmacist check the master formula for corrections and completeness.
3. Date and assign a batch number to the formulation and enter it in the records book.
4. Follow outline for compounding of solutions.

2. Checking Stock Solutions

1. Check cartons of commercial stock solutions received to ascertain that they are not damaged and are properly labeled.
2. Separate cartons of solutions which have water marks, unusual stains, breakage, or moldy growth, from the rest.
3. Open and examine defective cartons for cracked or broken containers, unusual color changes, defective labeling or other noticeable defects.
4. Send all damaged or defective cartons and solution bottles to the Receiving department to take corrective action in immediately returning and replacing the damaged and defective goods.
5. Store stock solutions in their appropriate place, making sure that the older solutions are brought forth to be used first.

3. Maintaining Adequate Space and Clean Facilities

1. Clean all counter tops with a detergent disinfectant followed by ethyl alcohol 95% prior to and after use.
2. Do not bring cartons or paper boxes into the compounding area. All stock solutions must be removed from their boxes and then transported on clean stainless-steel carts to the compounding area, where the outside of the containers is washed and rinsed (see Section II, C.3.a) before being shelved.
3. Do not crowd or overstock shelves.
4. Clean shelves and storage bins once a month and as required with a detergent disinfectant.
5. Clean floors daily with a detergent disinfectant.
6. Clean walls, ceilngs, and other room surfaces once a month with a detergent disinfectant.
7. Thoroughly rinse the stainless steel carts with hot tap water, then with distilled water, and towel-dry. Do this daily. Once a month wash top and bottom sides of carts with detergent. Carts must be stored dry to prevent rusting and the wheels oiled regularly (i.e., once a month).

4. Preparation of Equipment (Figure 9)

a. Preparation of a Hot Solution of 5% Trisodium Phosphate

1. Prepare a hot detergent solution of 5% trisodium phosphate by dissolving the appropriate amount of the salt in distilled water and boil. Use only when it is boiling hot.

b. Preparation of the Laminar Air-Flow Hood (LAFH)

1. Prior to use, shut off the UV germicidal lights in the LAFH.
2. Turn on the fluorescent lights and the LAFH operating switch.
3. Remove the protective plastic screen by folding neatly over the top of the LAFH.
4. Scrub the counter top and sides with cetrimide 1% followed by ethyl alcohol 95% and with great care also scrub the filter screen.
5. Allow the LAFH to operate for 30 min prior to use.

c. Removal of Pyrogens from Admixing Equipment

1. Thoroughly wash the stainless steel mixing tank, 142-mm Millipore® filter holder, required graduates, rubber stopper, pliers, hemostats, and scissors with the boiling trisodium phosphate 5% solution, then rise thoroughly with pyrogen-free distilled water.

d. Cleaning and Disinfecting Counter Tops, Peristaltic Pump, Top-Loading Balance, Retort Stand, Stainless-Steel Carts, and Tubing Holder

1. Scrub the stainless steel counter top of the filtration apparatus and the outside of the variable speed tubing pump with cetrimide 1% followed by ethyl alcohol 95%.
2. Scrub the pan and the entire surface of the top-loading balance with cetrimide 1% followed by ethyl alcohol 95%, then place carefully in the cleaned LAFH. Make sure the pan is locked during transport.
3. Scrub the retort stand as well as the tubing holder with cetrimide 1%, rinse with distilled water, then again with ethyl alcohol 95% and place inside the clean LAFH.
4. Wash the stainless steel carts thoroughly with distilled water and dry prior to use.

e. Rendering the Tygon® and Silicone Pressure Tubing Pyrogen Free

1. Remove used tubing from the Y-shaped glass connectors by placing under running hot tap water or boiling in tap water. Do not forcefully attempt to remove tubing from the Y connectors; this may result in breaking the glass connector with possible injury to the hand.
2. Remove defective portions of tubing by cutting them off with scissors.
3. Discard tubing that has become hard and brittle or deformed and cut new tubing.
4. Thoroughly rinse both the inside and the outside of the Tygon® or silicone tubing with hot tap water; then with distilled water.
5. Boil the tubing as well as a Teflon-coated sinking weight for 10 min in 5% trisodium phosphate solution.
6. Rinse thoroughly with pyrogen-free distilled water and place in a clean container.

f. Preparation of Sterile Tygon® Pressure Tubing Assembly

1. Remove used tubing from Y-shaped glass connector as described in Section III, B.4.e.
2. Discard damaged or brittle tubing and cut new tubing (3/8" O.D. and 3/16" I.D. diameters).
3. Thoroughly rinse the inside and outside of the used and precut Tygon® tubing as well as the glass Y connectors with hot tap water, then with distilled water.
4. Boil for 10 min in 5% trisodium phosphate, remove detergent water, and again bring to a boil in plain distilled water.
5. Rinse thoroughly with distilled water, then place in a clean container.
6. Assemble the 142-mm unit with a 0.22-μm membrane filter and then attach to the distilled water outlet.
7. Put on a gown, hair cap, mask, and gloves, and rinse thoroughly the inside and outside of the Tygon® pressure tubing; then place it in a clean container and cover with clean aluminum foil.
8. Carefully assemble the Tygon® pressure tubing unit under the LAFH; do not use excessive pressure when attaching tubing to the Y-shaped glass connector. This may cause the glass connector to break, causing injury.
9. Wrap tubing assembly unit with crepe autoclave paper by first separately wrapping each end of the tubing and then the entire assembly along its length with double wrapping. Do not overuse autoclave tape when securing the crepe paper.
10. Place in proper container and sterilize by autoclaving at 121°C for 21 min using wrapped cycle.

11. Remove sterilized units from the autoclave, stamp with a 3-month expiry date and store in appropriate area. Discard deformed or heavily discolored tubing.
12. Prepare such Tygon® pressure tubing assemblies in advance.

g. Assembly of 142-mm Millipore® Filter Holder

1. With great care attach the Teflon® O-rings to both the inlet and outlet plates of the 142-mm filter holder.
2. Place the underdrain support on the center part of the outlet plate making sure the smooth side of the underdrain support is exposed and that it is centered over the middle opening of the outlet plate.
3. Place the filter support screen over the underdrain support.
4. Place the membrane filter of choice over the support screen and wet with distilled water.
5. Place Dacron® separator (if required), then the other membrane filter of choice, and wet again with distilled water.
6. Place inlet plate and secure it in place with hand-wheel wrench. Tighten evenly but not excessively.
7. Place proper silicone gaskets at both triclover hose adaptors, then fasten securely using tri-clover clamps.
8. Take entire assembly into the sterile-filling room and place on table top beside the variable-speed pump.

h. Assembly and Connection of Silicone Tubing to the Variable-Speed Pump, Polypropylene Manifold, and to the 142-mm Millipore® Filter Unit

• Installation of Silicone Tubing to the Pump Head

1. Position the long depyrogenized silicone tubing (½" I.D.) around the rollers of the pump head and then pull to flatten. Gently work tubing and rotor assembly into pump-head bearing. Never use hard or sharp objects to seat tubing in place.
2. Position the free ends of the tubing in the inlet and outlet recesses of the pump-head housing and while continuing to exert tension on the tubing attach the other half of the pump head. It is necessary to maintain tension on the tubing to avoid improper seating of rotor assembly and damage to the drive unit during operation.
3. Make certain the tubing and rotor assembly are secure within the pump-head housing by rotating the assembly with a screwdriver while holding onto the housing.
4. Check that no gaps remain between the housing halves of the head pump and that no binding occurs between the tubing and the rollers.
5. Attach the assembled pump head to the drive unit using the two thumbscrews and washers. Make sure thumbscrews with washers are threaded into mounting holes opposite from those used previously.

• Connection of Silicone Tubing to the Polypropylene Manifold

1. Assemble the polypropylene manifold with the gauge, nipple, bypass valve, and threaded fittings.
2. Attach the end of the depyrogenized silicone tubing coming from the outlet recess of the pump-head housing to the inlet fitting of the polypropylene manifold and secure with a cable tie.
3. Attach a length of depyrogenized tubing to the bypass valve outlet to direct unfiltered fluid back to the mixing tank. Secure attachment with a cable tie.

• Connection of Tubing to the 142-mm Millipore® Filter Unit

1. Attach one end of the depyrogenized ½" I.D. silicone tubing to the outlet spout of the polypropylene manifold and the other to the inlet spout of the 142-mm Millipore® filter unit. Secure each connection with a cable tie.
2. Attach and secure with a cable tie one of the free ends of the depyrogenized Tygon® tubing (3/8" I.D. and approximately 24" long) to the outlet of the 142-mm filter unit.

i. Connection of Sterilizing Filter Assembly

1. Put on a gown, hair cap, and mask.
2. Scrub hands with chlorhexidine gluconate 2% or other appropriate skin cleanser, then put on gloves.
3. Carefully remove autoclave paper wrapping inside the LAFA and examine sterile tubing unit for holes or splits. Do not remove tubing end coverings.
4. Firmly secure all connections of the sterile tubing assembly with a cable tie.
5. Carefully remove the plastic protective cover from the sterile Twin-20 filter and attach to the metal retort stand. Do not remove the protective covering bell of the filter outlet. Check for any visible defects, such as damaged filter casing or broken filter. Do not use filter if there is any defect. Keep defective filter and lot number and return to manufacturer for replacement. Note that a loose metal clip in the center of the filter will not affect the integrity of the filter.
6. Connect the free end of the Tygon® tubing from the 142-mm filter holder outlet to the Twin-90 inlet. Secure with appropriate cable tie.
7. Carefully remove both the sterile tubing assembly inlet and Twin-90 outlet coverings and aseptically connect the free end of the tubing assembly to the filter outlet. Secure with appropriate cable tie.
8. Secure all three lengths of tubing connected to the Y connector with appropriate cable tie.
9. Support the two lengths of narrow tubing connected to the Y connector using a plastic test tube holder, then carefully remove the protective covering from the ends of the sterile tubing assembly and aseptically attach two sterile empty transfer packs.
10. Clamp one tubing length of the sterile tubing assembly by using a surgical vein clamp (hemostat).
11. Check to see that all connections are secure. Note that Steps 3 to 10 must be performed inside the LAFH.

5. Compounding of Solutions

a. Assembly and Preparation of Commercial Stock Solutions Prior to Compounding

1. Scrub hands thoroughly with soap prior to handling stock solutions.
2. Remove stock solution bottles from the cartons and check for defects such as cracks, unusual color, improper labeling, etc.; then place on a clean stainless-steel cart.
3. Take stock solutions into compounding area and rinse thoroughly with distilled water.
4. Towel-dry bottles; then sprinkle with ethyl alcohol 95%.
5. Have pharmacist check all stock solution bottles prior to opening.

b. Compounding

1. Close all doors to the sterile compounding admixing room. Allow no unauthorized personnel to enter the room.
2. Put on a gown, hair cap, and mask.
3. Scrub hands with a disinfectant detergent (e.g., chlorhexidine gluconate 2% skin cleanser).
4. Remove outer metal seals of the stock solution bottles, then the rubber diaphragm. If vacuum is absent do not use bottle.
5. Aseptically remove entire rubber closure using depyrogenized steel pliers.
6. Aseptically admix all components in the depryogenized mixing tank. *Note:* handle two bottle units at one time, and avoid splashing.
7. Mix well for 10 min using the electric bar mixer, then cover mixing tank with stainless cover and take into the sterile filling and admixing room.

c. Sterilization and Packaging of Solutions

1. Put on a gown, hair cap, and mask.
2. Scrub hands with a detergent disinfectant (e.g., chlorhexidine gluconate 2% skin cleanser) and put on gloves.
3. Tare balance using one of the empty transfer packs.
4. Turn the bypass valve adjustment knob counterclockwise until no resistance to rotation is detected. Do this prior to starting the pump.
5. Position the Twin-90 filter at a 45° angle and make sure one of the tubing outlets of the sterile tubing assembly is clamped off with a hemostat clamp.
6. Place one empty transfer pack on the balance.
7. Start the variable speed pump; then clamp the outlet tubing (½″ I.D.) between the manifold and 142-mm filter holder and turn the adjustment knob clockwise until the gauge registers a filtration pressure of 5 $lb/in.^2$
8. Remove all entrapped air by gently tapping the tubing until all of the surface area of the filter is completely covered with fluid.
9. Keep the first 200 g of solution for a sterility check.
10. Divert the filtration flow to the other empty transfer pack using the hemostat clamp.
11. Increase filtration pressure to 14 $lb/in.^2$ (no greater pressure at any time) by turning the adjustment knob on the manifold clockwise. Maintain that pressure at all times.
12. Fill each unit with the required weight of solution as stated on the master formula.
13. Seal each filled unit with a metal crimp and hand crimper.
14. Place filled unit on clean stainless steel cart.
16. Disconnect the spike of each full pack aseptically and with great care. If spillage occurs wipe off immediately with disinfectant, cetrimide 1%, and ethyl alcohol 95%.
17. Avoid contact with the sterile spike of each transfer pack and outlet openings of the sterile tubing assembly during the filtration process.
18. Obtain additional 4 × 200 g samples for sterility purposes during the filtration process.
19. Check integrity of Twin-90 filter when all the solution is filtered using the bubble point test at 14 $lb/in.^2$ Allow air to be pumped into the filter, displacing the remaining solution. If no bubbles of air are seen at the outlet of the Twin-90 at 14 $lb/in.^2$ the filter is intact. If bubbles are seen the batch must be filtered again. No solution must be released for use if the bubble point test fails.

20. Have two people working at all times during the filtration process, one to monitor the filling unit and the other to direct the flow of the filtrate.
21. Have the door of the sterile filling room closed at all times during filtration of the solution and at no time allow unauthorized persons to enter this area.
22. Do not bring any boxes, labels, or other unnecessary items into the sterile filling room.

d. Labeling and Packaging

1. Prepare labels containing batch number, ingredients, and where applicable, patient's name, volume, expiry date, and "refrigerate" auxiliary label.
2. Transport filled units of TPN solution into the prepackaging area.
3. Cut off excess tubing from each sterile unit using clean disinfected scissors.
4. Place filled unit on clean stainless-steel counter and label.
5. Place labeled units in clean plastic bags and seal with the thermosealer.
6. Package full finished units into clean cartons for home-care patients.

e. Storage

1. Store all solutions in the walk-in refrigerator, and enter into the record sheet.
2. Store solution on appropriate shelves.
3. Bring forth older solutions to be used first.
4. Do not overstack shelves to avoid breakage of plastic bags.
5. Remove all expired solutions and discard.

f. Transport to the Nursing Unit

1. Send all TPN solutions to the nursing unit by a pharmacy porter. If not feasible, make arrangements either with the floor pharmacist or nursing unit for transportation.
2. Instruct porters as to the critical storage condition and proper handling of these solutions.
3. Check returned solutions as to their storage condition prior to their delivery to the pharmacy, expiry date and other defects, before returning to stock. Improperly stored solutions, expired solutions or defective units must be discarded.

Chapter 16

QUALITY CONTROL

George Tsallas

TABLE OF CONTENTS

I. INTRODUCTION

Monitoring and assuring the quality and safety of the final product is an integral component of a TPN service and should be the responsibility of the pharmacy department.

The best way to be certain that a product contains the ingredients stated on the label is strict adherence to good manufacturing procedures. Simply performing tests cannot ensure the quality of a product which depends also on the ingredients used, the procedures utilized, and the personnel involved in the preparation of the product. The above aspects are of much greater importance in the preparation of solutions, particularly in a hospital setting where there is only limited time available for testing the product.

In the preparation of TPN solutions raw materials include commercial concentrates of amino acids, dextrose, electrolytes, and vitamins. It is imperative that commercial stock solutions be checked for proper labeling, expiry date, defects of the container, and unusual color changes of the solutions prior to use.

The efficiency and safety of equipment used in the preparation and packaging of TPN solution must be ensured.

II. LAMINAR AIR-FLOW HOOD (LAFH)

A critical piece of equipment is the LAFH. It is imperative that tests of it be carried out on a regular basis to be assured of safe operation of the unit. These tests should include the following.

A. Air Velocity

The velocity of air going through the high-efficiency particulate retention (HEPA) filter is measured using an anemometer or an air velocity meter. Air velocity is measured at frequent intervals across the face of the HEPA filter and in locations throughout the protected area. The air velocity measurements at the filter surface should match the manufacturer's specifications. Inoperative blowers, clogged prefilters, clogged HEPA filters, and an excessive leak in the air supply system all cause a reduction in air flow. Nonuniformity of air velocity may be due to poor air distribution in air supply and air return plenums, obstruction or offsets in filter bank, and nonuniformity of filters (usually because of the use of different brands of filters). The air velocity must always be within the specified limits of the manufacturer. Most LAFH are fitted with a static air-pressure gauge whereby an increase in pressure can be observed, indicating that a clogged HEPA filter requires changing. An air velocity below 70 ft/min indicates that the filter should be changed. In addition to the above-mentioned factors the presence of a leak in the HEPA filter can also drastically reduce air velocity, so that this finding also necessitates checking for leaks in the filter.

B. HEPA Filter Leak Test

A HEPA filter leak test should be performed during installation of the filter to be certain that the unit is properly functioning. This test should also be performed following change of the HEPA filter to detect failure due to defective gasket seals, clamping

FIGURE 1. Generator of dioctylphthalate (DOC) smoke (average particle size 0.3 μm) introduced into intake side of LAFH for testing filter retention.

devices, weld cracks, and other leaks that might develop during use. Nonoperational or standby units should also be tested because defects in these areas may take place even when these units are not being used.

The substance that is used to detect leaks is dioctylphthalate (DOP) smoke having an average particle size of 0.3 μm. This substance is introduced into the intake of the hood and into the upstream side of the HEPA filter. The concentration of DOP passing through the filter is measured using a smoke photometer (see Figures 1 and 2). Observation that no DOP is detectable indicates the filter is intact and that there are no leaks.

C. Air-Flow Patterns

Checking of air-flow patterns is of particular value in the training of personnel and in the proper utilization and placement of the LAFH. This test involves observation of smoke inside the working area of the hood. Smoke patterns (air-flow patterns) are essential in pointing out major areas of turbulence, dead spaces, and areas of back streaming. This very simple test is effective and inexpensive.

D. Proper Use of the LAFH

Ideally, the LAFH should run 24 hr a day. If a 24-hr service is not provided the LAFH must run at least 30 min prior to use.

The working counter top and sides of the LAFH must be thoroughly cleaned with an appropriate disinfectant prior to and after each use.

Bottles and other containers brought inside the LAFH must be cleansed prior to placing inside the LAFH to minimize the introduction of microbial contamination and particulate matter.

Cartons of intravenous solutions must not be opened in the LAFH nor even taken inside.

Objects inside the LAFH must be arranged in such a manner as to allow sufficient space between them for a good flow of filtered air.

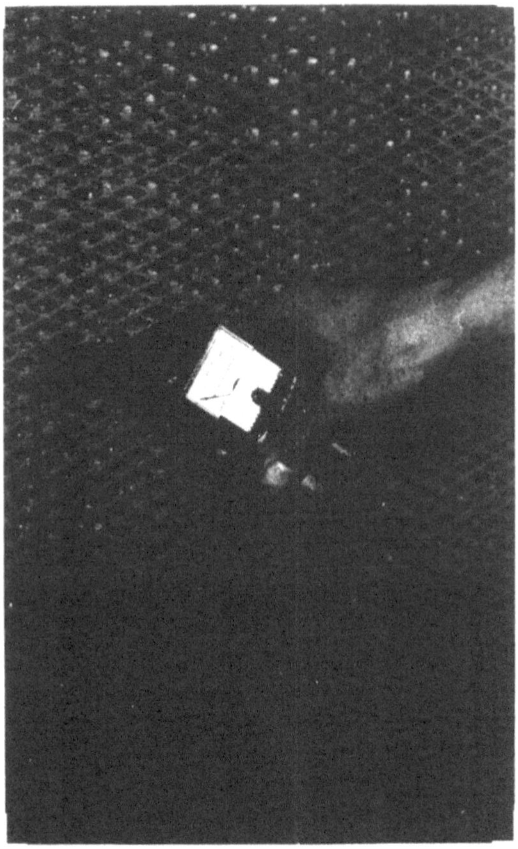

FIGURE 2. Detector on output side of LAFH for DOC smoke that may have escaped entrapment in the HEPA filter. Complete entrapment is a measure of filter integrity.

Work flow must be arranged and performed in such a manner as to minimize turbulence and to prevent clean filtered air from washing over dirty articles and contaminating other articles that must remain clean.

Personnel should wear gowns, mask, gloves, and hair caps during extensive manipulation while compounding, and must refrain from unnecessary conversation, movements or interruptions when working in the LAFH.

Suitable ultraviolet light should be used inside the LAFH. There should be a protective curtain that can be drawn when the facility is not in use.

Work should be carried out 6 in. within the hood and no elbows, arms, or hands should rest on the working surface of the hood.

Recording or calculations must not be carried out on the working bench of the LAFH.

E. Microbiological Testing

Although there are no formal guidelines or procedures with regard to microbial contamination monitoring of the LAFH, microbiological testing can be a very useful procedure in monitoring the working environment of the LAFH. There is no consistency in the procedures used between institutions. Each establishes its own method of microbiological testing. Generally the method involves both microbiological air sampling and work-surface sampling. The first method involves the use of a mechanical air sampler where the LAFH air passes through a sterile membrane filter that is cultured to look for organisms in the atmosphere. The second method involves sampling of the work surface then culturing, or the use of the Rodac® plate technique.

A more practical and commonly used method (although it is not very reliable) for microbiologically monitoring the LAFH environment is to expose blood-agar plates inside the work area of the LAFH. Usually six to eight blood-agar plates are set up in a zig-zag fashion and exposed to the LAFH environment over a period of 2 hr. The plates are next incubated and observed for growth. This procedure should be done about once a month and the results recorded. If growth occurs the organism should be identified, the hood resampled, and an investigation should be done for possible correlation with routine TPN compounding. If a second sample is still unacceptable the technique of the personnel should be evaluated. Further unsatisfactory sampling requires LAFH testing by a commercial agency.

The swab test and agar plate surface-contact method (i.e., Rodac® plate) can also be used despite limited hospital facilities. It is important however to realize that chemicals on the surface, such as a disinfectant, may interfere with the results. There are commercial swab testing kits that can also be used to monitor the total bacteria count on work surfaces.

Although great emphasis is placed upon the HEPA filter of the LAFH it is also important to consider the efficiency of the prefilters of the LAFH. These filters must be changed once every 1 to 3 months, depending on the use of the LAFH and the environment in which the hood operates.

Some of the tests discussed such as microbiological monitoring and air-flow patterns can be carried out by the pharmacist. Tests for leaks and air velocity of the HEPA filter should be done once every six months by a reputable firm.

Additional equipment such as mixing tanks and filtration systems used in the preparation of solutions must be kept clean at all times and in good operating order. Furthermore, such equipment should be restricted for use to the manufacture of sterile products.

III. FACILITIES

The area where solutions are prepared should be cleaned daily and, where feasible, microbiologically monitored (i.e., air sampling and/or work-surface sampling). Ceilings, walls, and floors should be cleaned with a disinfectant on a regular basis and air filters changed to ensure good quality of air. Handwashing facilities and work benches should also be kept clean at all times.

IV. PERSONNEL

The proficiency of personnel in all areas of solution manufacture should be monitored for correct technique and adherence to procedural guidelines. Disciplinary action should be taken with personnel not following the protocol. In-service education programs should be implemented to formally train technical personnel and update their technique. Monitoring of technique at predetermined intervals should be done with all personnel involved in the preparation of solutions in order to validate performance.

V. THE END PRODUCT OR FINISHED PRODUCT

The finished product must be tested for freedom from microbes, pyrogens, endotoxins, and particulate matter. If feasible a quantitative analysis should be done to monitor the concentration of admixed components.

A. Sterility

Sterility testing may be done by random selection of samples of the product. Usually they can be carried out by the microbiological department of the institution. Presently there are commercially available systems such as the Addi-Chek® (Millipore, Bedford, Mass.) quality control system that can be used to sample solutions for microbial contamination. It is basically a membrane filter through which a full unit of solution is introduced into a closed clear plastic canister with a bacterial retentive filter (0.45 μm) and a hydrophobic air vent at the top of the canister. The filtered solution is discarded, the outlet of the canister is sealed, and growth medium is introduced into the canister by gravity. The canister is labeled then incubated at 20 to 25°C. Microbial contamination is indicated by the presence of turbidity within a 10-day period. A positive culture can be analyzed for the type of contamination using standard microbiological procedures. The system is very useful since a large volume of sample or an entire admixture unit can be tested. The system will be particulary useful with the system of bulk compounding of TPN solutions. It is, however, expensive and thus limited in its use.

Sterility testing is a more useful procedure in validating the actual compounding process than it is for establishing acceptability of the units manufactured. It is really a check on the program. Although there are various sampling systems, random sampling is still the most common and most practical, but not statistically rigorous.

B. Pyrogen Testing

This test is not always practical, mainly because of lack of facilities and of high cost. However for solutions that are routinely compounded in bulk and stored for future use pyrogen testing is essential to ensure that pyrogen agents are not introduced into the solution by the equipment used or by the procedure.

Although the in vivo rabbit test is the official test for pyrogens, the in vitro Limulus test is very effective and more sensitive in detecting the presence of bacterial endotoxin.

C. The Limulus Amoebocyte Lysate (LAL) Test for Bacterial Endotoxin

1. General

The LAL test is a fairly recent development in the area of microbial contamination. It is an in vitro test based on the intravascular clotting mechanism of the injected horseshoe crab, *Limulus polyphenus.* Upon exposure to extremely low concentrations of endotoxin special blood cells or amoebocytes of the crab aggregate to form a loose network and a protein gel due to the reaction between the endotoxin and clottable proteins of these blood cells. These amoebocytes presently are refined by concentrating, washing, lysing, and lyophilizing procedures, which leave these clottable proteins in a suitably powdered form that becomes reactive upon dissolution when it may again be used to detect bacterial endotoxin. The mechanism of the LAL reaction (see Figure 3) appears to involve the presence of an inactive proenzyme which becomes activated in the presence of endotoxin or a suitably prepared lipid-A derivative of endotoxin, plus as well as the presence of metal cations such as Ca^{2+} and Mn^{2+}. This activated enzyme has been characterized as belonging to the serine protease class and is similar to other enzymes of this type such as thrombin, trypsin, and factor XA. The activated enzyme then reacts with a lower molecular weight protein fraction of the LAL called coagulogen. The coagulogen is cleaved by the enzyme into at least two subunits, one of which remains soluble. The other coagulogen fragment(s) interacts physically and/or chemically, ultimately resulting in a visible, solid gel, a flocculent precipitate, or in turbidity, depending upon the design of the test system.

The LAL test is the most sensitive and convenient method available for the detection of bacterial endotoxin. In comparison with the rabbit test it offers several advantages.

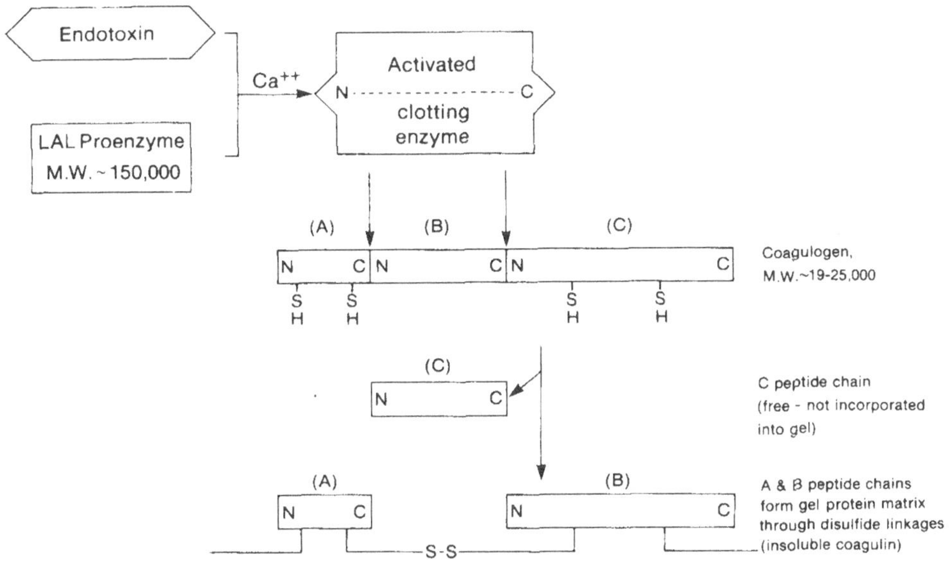

FIGURE 3. Schematic representation of LAL gel reaction mechanism. *Note:* C———N = C and N terminals of peptide chain. (A), (B), (C) refer to peptide constituents in coagulogen. (A) and (B) only form gel.

It is more sensitive, reliable, specific, reproducible, and economical, with wider applicability. U.S. regulatory agencies have now recognized the LAL test to be an alternative to the USP rabbit test.

There are, however, certain factors that may interfere with the test. These include heavy metals, surfactants, pH, and characteristics of the endotoxin itself. These factors may interfere with or enhance the LAL reaction. However, through full understanding of the LAL reaction mechanism appropriate adaptive measures can be taken to overcome these problems. For example, the degree of interference of these agents may be determined by comparing endotoxin dilutions in the presence and absence of the interfering substance. Solutions with either very acidic or basic pH should have the pH adjusted to neutral prior to eliciting the reaction.

Although other modifications of the LAL test have been developed for estimating the relative amount or concentration of endotoxin — such as measuring the degree of turbidity or the amount of precipitated coagulogen — the qualitative approach of checking for a firm clot is still the most commonly used method, being simple, reliable, and cheap. Furthermore, there seems to be some disagreement amongst experienced users as to whether instrumental methods of quantitative analysis offer enough improvement over the gel method, in terms of precision, to merit their higher cost and complexity.

Basically the test involves combining the LAL reagent with an equal volume of sample solution and incubating the mixture for a period of 1 to 4 hr at 37°C. A positive end-point is recorded when a visible solid gel capable of maintaining its integrity when the test tube is inverted 180° is observed. A positive control is used to ensure the system is functioning. Such a result indicates the presence of sufficient endotoxin to elicit a pyrogenic response if injected into a patient. More recently, as mentioned above, the end-point has been measured instrumentally as the degree of viscosity or as the amount of protein precipitated (by optical density or light scattering) against a reference standard. Such refinements have greatly increased the sensitivity of the test and have allowed quantification of endotoxin. As noted above there is, however, debate over the clinical significance of this refinement of the measurement.

TORONTO GENERAL HOSPITAL - PHARMACY DEPARTMENT

LIMULUS AMEBOCYTE LYSATE TEST

Performed By____________ Date__________ Pyrogen Test No.__________

Lysate Used: __________ __________ __________ Date__________

Brand Name Manufacturer Lot No. Lysate Prepared

Sample	Manuf.	Lot #	Reaction Time (Hrs)				Results		Comments
			0.5	1.0	2.0	4.0	Passed	Failed	
Pc									
Nc									
S_1									
S_2									
S_3									
S_4									
S_5									
S_6									

Pc: Positive Control (Endotoxin Reference Solution).
Nc: Negative Control (Sterile Water for Inj. without bacteriostat).
S_1-S_6: Samples to be tested.

Volumes - Sample: 0.1 ml (including controls), Lysate: 0.1 ml. Reaction Temp.: 37°C

Recording Results:

1. Positive: +++ = SOLID Gel
   ++ = Weak Gel
   + = No Gel (but increased viscosity, opacity or starchy granules)
2. Negative: - = No Gel (flocculation is regarded as negative)

Interpretation:

PYROGENIC: SOLID Gel within 4 hours

Non-Pyrogenic: No Gel within 4 hours

pH Adjustment (if any)

____

____

____

Supervisor:____________________

Date:____________________

FIGURE 4. Toronto General Hospital Pharmacy Department record form for Limulus amoebocyte lysate test.

2. Detailed Procedure

The following is a detailed procedure of the single LAL gel test used by the Toronto General Hospital pharmacy to examine for bacterial endotoxin in TPN and related manufactured sterile solutions (see Figure 4).

a. Materials and Equipment

E-Toxate® reagent and reference endotoxin (both Sigma Laboratories, St. Louis, Mo.)
10 × 75 mm Glass test tubes
1.0 m*ℓ* Graduated disposable pipettes
400 m*ℓ* Tall Pyrex® glass beaker
1-m*ℓ* Sterile disposable syringes with 25-gauge needles
Aluminum foil

Parafilm® (American Can Co., Dixie/Marathon, Greenwich, Conn.)
50 mℓ Glass beakers
Hot-air oven
Test-tube stand
Masking tape
Laminar air-flow hood (LAFH)
Water bath
Centrifuge
Sterile water for injection (without bacteriostatic agent)

b. Preparation of Glassware Equipment

1. Place sufficient dichromic acid in the 400-mℓ glass beaker.
2. Carefully immerse the reaction test tubes in the acid and soak overnight.
3. Rinse the test tubes thoroughly and individually with purified water and finally with pyrogen-free water (i.e., sterile water for injection).
4. Thoroughly rinse the 400-mℓ glass beaker with purified water and finally with pyrogen-free water.
5. Place the drained test tubes, inverted, in the 400-mℓ glass beaker.
6. Thoroughly rinse a piece of aluminum foil and use it to tightly cover the beaker with the test tubes.
7. Thoroughly rinse the pipettes with purified water and pyrogen-free water.
8. Drain the pipettes and then wrap well in aluminum foil previously rinsed with pyrogen-free water.
9. Cut a sufficient number of aluminum squares for covering the test tubes, rinse with pyrogen-free water, then wrap in a larger piece of aluminum foil previously rinsed with pyrogen-free water.

c. Heat Sterilization of Glassware

1. Turn on hot-air oven and set to 250°C.
2. Place wrapped test tubes, pipettes and aluminum foil in hot-air oven and heat at 250°C for 90 min.

d. Setting up of Water Bath

1. Fill water bath to appropriate level with distilled water.
2. Heat water to 37 ± 1°C and maintain this temperature.

e. Preparation and Labeling of Test Tube Stand

1. Wash aluminum test tube stand with detergent, then rinse thoroughly with distilled water.
2. Rinse again with ethyl alcohol 95%.
3. Prepare code labels using masking or autoclave tape, as on the pyrogen test record form, and affix appropriately to test tube stand. Prepare another set of similar code labels for test tubes.
4. Place test-tube stand in water bath.

f. Preparation of the LAFH

1. Turn on LAFH.
2. Scrub working surface first with cetrimide 1% followed by ethyl alcohol 95%.
3. Let LAFH operate for 30 min prior to use.

g. Preparation of Solution Samples for Endotoxin Detection

1. Remove TPN solution samples from the refrigerator and place inside the LAFH.
2. Aseptically attach sample coupler injection sites and let solution come to room temperature.

h. Preparation of LAL Reagent and Reference Endotoxin

1. Remove aluminum seal from LAL reagent vial.
2. Aseptically add 2.5 m*l* of sterile water for injection that does not contain a bacteriostatic agent, then swirl to dissolve.
3. Using a dry pyrogen-free pipette, transfer contents into an appropriate size of pyrogen-free glass test tube.
4. Cap with pyrogen-free aluminum foil and centrifuge at 1000 to 2000 g for 10 min.
5. Remove the test tube with the LAL reagent from centrifuge and place inside the LAFH.
6. If frozen, reconstituted LAL is available, thaw and if necessary centrifuge, as in Step 4 above.
7. Place reference endotoxin inside the LAFH.

i. Setting Up of Sterile Pyrogen-Free Glassware

1. Place sterilized glassware inside the LAFH.
2. Allow glassware to cool down to room temperature prior to use.
3. Do not remove aluminum foil cover from any glassware until ready to use.

j. Setting Up and Labeling of Glass Test Tubes

1. Put on gloves.
2. Carefully remove glass test tubes from glass beaker, label and place in designated spot in test-tube holder.

k. LAL Reaction Procedure

1. Aseptically, without stirring up the LAL reagent, draw up 1 m*l* of it.
2. Aseptically transfer 0.1-m*l* aliquots of the LAL reagent to each test tube.
3. Using another sterile pipette draw 0.1 m*l* of reference endotoxin and carefully transfer it into a test tube labeled PC (positive control). Swirl contents slowly to thoroughly mix endotoxin and LAL reagent.
4. Cover test tube with sterile aluminum foil and then seal with Parafilm®. Place it in the test-tube holder in the water bath to incubate at 37°C.
5. Using still another pipette draw up 0.1 m*l* of sterile water for injection and transfer it to a further test tube of LAL reagent labeled NC (negative control). As in Step 4 swirl contents to effect mixing, then cover with sterile aluminum foil and again with Parafilm®. Next place test tube in water bath at appropriate (labeled) point in rack, to incubate at 37°C.
6. Similarly, as in Steps 4 and 5, using separate pipettes or syringes for every sample transfer 0.1 m*l* of the product samples to appropriate test tubes and incubate at 37°C.
7. Record data required for pyrogen test form and note time of incubation for each sample.

8. Observe for gel formation at ½, 1, 2, and 4 hr.
9. Using the notation listed on the form for interpreting results, note the observations requested on the pyrogen-test form.
10. Have results checked by a pharmacist.
11. Doubtful readings must be repeated.
12. Place reference endotoxin back in the refrigerator and excess reconstituted LAL reagent in the freezer. Date of reconstitution must be stated on the LAL reagent.

I. Precautions

1. Avoid touch contamination during transferring process.
2. Control temperature of water bath within required range.
3. Do not use shaking to admix the LAL and samples because shaking causes gel formation, but mix with swirling.
4. Adjust pH for solutions not close to neutral pH.
5. When taking visual readings do not shake reacting tubes. Slowly tilt tubes to an approximately 180° angle and compare the gel formation with that in a positive control.
6. When reacting LAL reagent and samples in test tubes do so slowly in order to avoid bubble formation during sample transfer from pipette or syringe.

D. Particulate Matter

All solutions should be visually examined for particulate matter. This test for clarity usually involves searching for a Tyndall phenomenon, viewing the container by tranverse light against light and dark backgrounds (for dark- and light-colored particles, respectively). Solutions showing particulate matter should be discarded or refiltered. Although there are other more sophisticated methods the visual method is the most practical.

E. Quantitative Analysis

Finished products that are stored for future use should be quantitatively analyzed. Some quantitative determinations such as amino acid profile, glucose content, various electrolyte concentrations, pH, and osmolality may be determined by the hospital biochemistry laboratory if departmental facilities are not available or if outside laboratory costs are prohibitive. Because of unavailable facilities or prohibitive costs it may not always be feasible to regularly test quantitatively for certain components of the admixture, such as amino acid or dextrose content, and this is acceptable with well-established standard preparations.

It is essential that proper documentation be kept for all quality control procedures. Periodically, such records should be reviewed and evaluated.

VI. SUGGESTED GUIDELINES FOR QUALITY CONTROL

Each pharmacy department should have written policies and procedures for quality control which are continually reviewed and evaluated. These give maximum assurance of the quality of the product in terms of sterility, apyrogenicity, clarity, freedom from particulate matter, and potency when feasible.

A. Equipment

LAFH units should be inspected and certified at regular intervals to ensure efficacy and safety of the unit. The following parameters should be tested according to the instruction manual for the LAFH. This should be carried out by the supplier of the LAFH or by a reputable testing firm every six months.

1. Air velocity
2. Air-flow patterns
3. Filtration (DOP) testing

The LAFH should be routinely monitored microbiologically (i.e., once a month) and the prefilter(s) changed every 1 to 3 months as required.

It is also desirable, if an ultraviolet light is not part of the original equipment, that a portable one be activated inside the hood when it is not in use and a protective curtain drawn across its front.

B. Personnel

It is the responsibility of the pharmacist to monitor and evaluate the aseptic technique used by personnel who are compounding TPN solutions.

C. Product

Representative samples of the solutions prepared should be checked on a regular basis for sterility to ensure the safety of the product. If the need arises, the official USP pyrogen test should be performed. For practical purposes the LAL pyrogen test may be used for in-process quality control.

A permanent quality control record should be kept of all samples sent for sterility and pyrogenicity testing. These records should be readily accessible and reviewed periodically.

Compounded products should be examined for particulate matter and clarity over a projected beam of light against a black background for light-colored particles and a white background for dark-colored particles (as noted earlier in Section V, D).

Glass bottles should be examined for cracks and solution bags squeezed to test for leaks. All containers should be checked to ensure that they are properly sealed prior to distribution. Adequate controls should be included in the compounding process to assure one of the potency of the final product.

It is advisable to follow manufacturer's recommendations regarding admixing and storage of solutions.

The pharmacist must check the label on the finished product for patient identity, location, medication and amount added, type and volume of solutions, accuracy of expiration date, and storage directions.

The pharmacist must also check the order against additives and amounts actually used by examination of empty vials, ampules, used syringes, and volume of solution used.

VII. AN ILLUSTRATIVE SET OF QUALITY CONTROL PROCEDURES

The following is an outline of the quality control procedures used by the Pharmaceutical Services of the Toronto General Hospital to monitor the preparation of TPN solutions.

A. Laminar Air Flow Hood (LAFH)

1. Testing

The LAFH will be tested once a year by a reputable firm for the following:

1. Air velocity
2. Air flow patterns
3. Dioctylpthalate smoke test

2. Change of Prefilter

1. Change prefilter once a month.
2. Record the date of change in control record books.

3. Microbiological Test of the LAFH

The microbiological test will be performed once every four weeks by qualified personnel.

4. Plate Culture Test

Blood agar plates will be positioned, six in each LAFH when the hoods are in operations, then sent to microbiology laboratories for culturing. This procedure will be done as follows:

1. Gown, cap and wash hands, prior to engaging in the procedure.
2. Carefully remove covers of plates inside the LAFH and position the plates approximately 12″ apart in a zig-zag fashion on the working bench surface.
3. Leave plates for 120 minutes while hood is in operation.
4. Carefully cover plates, and label; then fill out the proper microbiology requisitions and identify time of exposure, LAFH, date test performed, and deliver to microbiology laboratory.
5. Perform this procedure for each LAFH.
6. Record date test performed and results in appropriate control record book.

B. Other Equipment

Periodically check the peristaltic pump, mixing tank, and filtering apparatus for cleanliness and maintenance as required.

C. Facilities

1. Sterile Compounding and Filling Rooms

- Scrub all working counter tops daily with a detergent.
- Once a month disinfect walls and ceiling of the sterile compounding and filling area with a detergent disinfectant. Record date.

2. Distilled Water Supply

- Change UV lamp of distilled water storage tank every 90 days or as needed. Record date of change.
- Once a month send samples from all distilled water outlets including the storage tank to the Microbiology department and also test for presence of pyrogens using the LAL test.

D. Solutions

1. Commercial Products

1. Examine cartons of solutions for any stains or damages.
2. Examine solutions for clarity, unusual color change, cracked bottles, particulate matter, and proper labeling.

2. Finished Product

Sterility Testing of Solutions Prepared by the Bulk Method

- Take five samples (approximately 300 mℓ in size) during the sterile filtration of the product.
- Fill out microbiology requisition form for every sample.
- Send three samples (first, last, and one in between) to microbiology, following completion of the batch.
- Keep two samples in the refrigerator and send one for culture after one month and another after two months.
- Identify requisition results with type of product, using the manufacturing record book, and enter in sterility control record book; then attach to the appropriate master formula. Report positive results to the supervisor. Positive results must be followed up with microbiology. Product must not be released unless cleared by microbiology. In the event that the product has been released it must be recalled.

Biochemical Analysis for Solutions Prepared by the Bulk Method

- Send a sample (approximately 20 mℓ) to the Biochemistry department with appropriate requisitions.
- Match requisition results with type of product, using manufacturing control record book.
- Attach results to master formula and enter results in record book.
- Discuss discrepancies with supervisor.

Sterility Testing of Solutions Prepared by the Individual Bottle Method

- Take at random one sample (e.g., 25 mℓ) in a sterile eye-dropper bottle for every three units prepared.
- Fill out microbiology requisition form for every sample and send for culturing.
- Match requisition results with type of product using, the manufacturing record book, and enter in sterility control record book. Then attach requisition results to the appropriate master formula. Report any positive results to the supervisor.

Pyrogen Testing

Pyrogen testing need be performed only with TPN solutions prepared by the bulk method (e.g., standard and homecare).

- Prepare all necessary equipment as described in pyrogen test procedure (see Section V, C).
- Take a pyrogen-test record form and fill out as directed (Figure 5).
- Take TPN solution sample from the refrigerator, identify lot number on pyrogen test record form, then place inside the LAFH.
- Wash hands thoroughly, glove, gown, cap, and mask.
- Insert site couplers (injection sites) to the various TPN solution units.
- Proceed with the test as described in pyrogen test procedure (Section V, C).
- Record results on pyrogen-test record form.
- Have pharmacist check results prior to completion of the test.
- Have supervisor check and sign record form prior to filing.
- Make necessary entry of results in appropriate record book and on the master formula.

TORONTO GENERAL HOSPITAL PHARMACY DEPARTMENT

DIVISION OF MANUFACTURING

MICROBIOLOGICAL AND PYROGEN TEST RECORDS

Date Sample Sent/ Tested	Number of Samples	Batch Number	Preparation or Item Tested	Initials	Date Results Received or Obtained	Micro-biology Lab No.	Pyrogen Test Number	RESULTS				REMARKS	Initials
								Microbiology		Pyrogen Test			
								+ve	-ve	Pass	Fail		

FIGURE 5A

FIGURE 5. Replica of forms used at the Toronto General Hospital for recording results of microbiological and pyrogen testing of nutrient solutions (A) with biochemistry record (B).

TORONTO GENERAL HOSPITAL PHARMACY DEPARTMENT

DIVISION OF MANUFACTURING

BIOCHEMISTRY RECORD

Date Sample Sent	No. of Samples Sent	Batch Number	Preparation	Initials	Date Results Received	Biochem. Lab Number	Results		Remarks	Initials
							Satis-factory	Unsatis-factory		

FIGURE 5B

Particulate Matter Check

- Check all finished units visually for the presence of particulate matter.
- Reject any unit with visible particulate matter.

Leak Test

- Squeeze the plastic bag units for possible leaks either due to improper sealing of units or manufacturer's defect.
- Reject any units leaking or found improperly sealed.

E. Sterile Manufacture and Filling Room Control Standards

1. Room Standards

Physical Plant

1. Doors — must remain closed at all times.
2. Floors — must be cleaned daily and be free of litter.
3. Ceilings — must be cleaned once a month with a detergent disinfectant.
4. Walls — must be cleaned once a month with a detergent disinfectant.
5. Air vent filters — must be changed once a month.
6. Sinks and counter tops — must be cleaned thoroughly at the end of the day.

Equipment

1. LAFH — must operate at all times during the day, with the curtain drawn over the opening and the UV light on during the night.
2. LAFH working surfaces — must be cleaned every morning, before use, and at the end of the day as described earlier.
3. Mixing tank and filtering system — must be cleaned thoroughly prior to and after use as described by method.
4. Stainless steel carts — must be cleaned prior to use and at the end of the day, dried, and stored properly.
5. Graduates — must be cleaned prior to and after use and stored properly.
6. Mixing bars, clamps, and decaping pliers — must be cleaned prior to and after use and stored properly.

2. Personnel Standards

Clothing

1. Uniforms — must wear gown, cap, mask, and glove during the manufacture and sterile handling of solutions.
2. Restrictions — must wear uniform only in the manufacturing and sterile filling room areas.

Conduct

1. Food and drinks — no food is permitted in the sterile manufacturing and filling rooms.
2. Smoking — no smoking is permitted in the sterile manufacturing and filling room.
3. Socializing — no socializing is permitted in the sterile manufacturing and filling room.
4. Personal articles — no items such as lipstick, powders, etc. will be used in the sterile manufacturing and filling rooms.

3. Work Practices

1. All intravenous admixtures — must be done inside the LAFH and within the appropriate area.
2. Manufacture of extemporaneous products — must be done according to "Master Formula" procedures.

READING LIST

Laminar Air Flow Hood and Environmental Control

Davies, W. L., Lamy, P. P., and Kitler, M. E., Environmental control with laminar flow, *Hosp. Pharm.*, 4, 8, 1969.

Gross, R. I., Laminar flow equipment: performance and testing requirements, *Bull Parent. Drug Assoc.*, 30, 143, 1976.

Lamy, P. P., Davies, W. L., and Kitler, M. E., Contamination control with laminar flow hoods, *Hosp. Pharm.*, 3, 12, 1968.

Phillips, G. B. and Runkle, R. S., *Biomedical Applications of Laminar Air Flow*, CRC Press, Cleveland, 1973, 37.

The Limulus Amoebocyte Lysate (LAL) Test

Cooper, J. F., Principles and applications of the limulus test for pyrogens on parenteral drugs, *Bull. Parent. Drug Assoc.*, 29, 122, 1979.

Cooper, J. F., and Neely, M.E., Validation of the LAL test for end-product evaluation, *Pharm. Technol.*, 4, 72, 1980.

Maxoli, C. and Weary, M., LAL test for detecting pyrogens in parenteral injectables products and medical devices. Advantages for manufacturers and regulatory officials, *J. Parent. Drug Assoc.*, 33, 81, 1979.

Pyrogen Monograph, Mallinckrodt Inc., St. Louis, Mo., 1978.

Weary, M. and Baker, B., Utilization of the limulus amebocyte lysate test for pyrogen testing large volume parenterals, administration sets and medical devices, *Bull. Parent. Drug Assoc.*, 31, 127, 1977.

Chapter 17

STABILITY AND COMPATIBILITY OF TPN ADMIXTURES

George Tsallas

TABLE OF CONTENTS

I. INTRODUCTION

Total parenteral nutrition (TPN) admixtures are very complex multicomponent solutions with great potential for physicochemical interactions that may be extremely difficult to analyze.

There is relatively little information available on the stability and compatibility of TPN admixtures. A primary reason for this lack of information is the difficulty in carrying out stability studies of TPN admixtures: a result of their complexity and often the need for sophisticated instrumentation. Most of the information available on incompatibilities centers about the physical changes in the final admixture, such as clarity, color, and precipitation.

II. COMPATIBILITY AND STABILITY OF AMINO ACIDS WITH DEXTROSE

Commercially available amino acid solutions are quite stable for periods exceeding a year. Dextrose is the most common carbohydrate mixed with amino acids in TPN solutions in North America. The usual reaction between dextrose and amino acids is the "browning effect" or Maillard reaction. This reaction is affected by both the temperature and the pH of the solution. TPN solutions stored at 4°C for periods of up to 12 weeks appear to be stable and develop no color. The bisulfite used as a preservative in certain amino acid solutions has been reported by Kleiman et al. to cause a 20% loss of tryptophan in solutions stored at 25°C for a period of 4 hr. The packaging of amino acid solutions in amber glass bottles reduced the degradation of tryptophan to only 4% over a period of 6 months at 25°C. More recent work by Farago has shown that even at 4°C, over a period of 12 weeks there is an average loss of 33% of the tryptophan present in a TPN admixture containing 4.25% amino acids and 16.65% dextrose. In the presence of electrolytes and trace metals the loss increased to 44% of the available tryptophan. On continued storage the lower concentration of tryptophan subsequently remained virtually unchanged from the 14th to 21st day of storage. This finding was in agreement with the work of Kleinman, who concluded that the decay of tryptophan is exponential and that no further degradation occurs after 15 days.

Studies performed by the author have shown that amino acid solutions with dextrose and minerals show no significant changes in the amino acid profile of the TPN admixture when stored for 8 weeks at 4°C. It must be remembered, however, that the method of manufacture, the amount of exposure to oxygen, light, and heat, and the concentration of other components will affect the stability of the amino acid admixture.

III. STABILITY AND COMPATIBILITY OF ELEMENTS ADDED TO TPN SOLUTIONS

A. Electrolytes

1. General

Most mixtures of amino acids and dextrose per se do not contain sufficient electrolytes to meet the needs of the patient. Therefore, additional electrolytes must be added to the basic TPN admixture. This addition of electrolytes to TPN admixtures may result in undesirable reactions with other components of the solution.

The most commonly seen reaction of this sort occurs between calcium and phosphate in the TPN mixture. Calcium phosphate precipitation has been reported following addition of calcium with phosphate to a TPN solution. The precipitate or crystal consists mainly of a mixture of monobasic and dibasic calcium phosphate depending upon the concentration of the reactants, the temperature, and pH. At higher temperatures precipitation occurs faster than in refrigerated solutions. Also lower pH is more likely to keep the calcium and phosphate in solution. In addition to the above factors it has been shown that the type of calcium salt and order of mixing seem to influence the occurrence and rate of precipitation. Studies done by the author have shown that the addition of 44.7 meq of calcium gluconate or a gluconogalactogluconate salt to a 3% amino acid solution with or without 17% dextrose, but with 24 meq of phosphate and 11 meq of magnesium sulfate, did not result in a precipitate when observed for a period of 60 days at 4°C. By contrast, only 11 meq of calcium chloride could be added to the same amount of phosphate and magnesium sulfate without precipitating. The ease with which calcium chloride precipitates may be due to the greater degree of ionization of calcium chloride, making calcium ions more readily available for interaction with phosphate and sulfate anions. A more complex and poorly ionized calcium salt such as gluconate may not interact easily with phosphate. In addition, the order of mixing of these salts can also alter the occurrence and rate of precipitation of the calcium-phosphate complex. Adding calcium as the final ingredient in a TPN admixture (after addition of the phosphate) also reduces the likelihood of interaction. Nedich has examined and reported the effect of pH, calcium, and phosphate concentrations on the precipitation of calcium-phosphate complex in Travasol®-Dextrose admixture. He ascertained a curve of pH to calcium concentration that defined the zone of precipitation (Figure 1). In addition Nedich also defined the relationship of phosphate and calcium concentrations in Travasol® which also delineated a zone of precipitation. Admixtures with calcium-phosphate ratios falling below the defined curve (Figure 2) did not show precipitation. Hence to avoid precipitation the pH, calcium, and phosphorous concentrations must be set so as to fall outside the zone of precipitation defined by the curves.

Additionally, a potential incompatibility with calcium is the interaction with bicarbonate when it is added to TPN mixtures. Such incompatibility can result either from a pH change or from the precipitation of insoluble calcium bicarbonate. Although bicarbonate may be added to TPN admixtures containing calcium, the occurrence of interaction is concentration-dependent. In addition the presence of other potentially interacting anions will also influence compatibility. To avoid the problems it is advisable that another salt such as sodium acetate or lactate be used to generate bicarbonate metabolically in vivo instead of adding bicarbonate to the solution.

2. Recommendations for the Addition of Electrolytes to TPN Solutions

1. Monovalent electrolytes such as sodium and potassium should be added first.
2. Phosphate should be added prior to the addition of calcium salts.
3. Addition of bicarbonate should be avoided. Instead substitute with acetate and lactate.
4. pH adjustment of the TPN solution should be avoided or else kept within the range of calcium and phosphate solubility.
5. A ratio of calcium to phosphate of 1:2 can be used as a rule of thumb if the two are admixed together.
6. Consideration should be given to electrolytes already present in the solution.
7. Consideration should be given to calcium-phosphate delayed crystal formation during storage of the TPN admixture.
8. Addition of two incompatible ions such as calcium and phosphate may be done separately into only one of the basic TPN components. Calcium may be added to the carbohydrate solution and the phosphate to the amino acid solution prior to admixing of the two solutions.
9. Thorough mixing is essential after the addition of each electrolyte.
10. Calcium chloride should be avoided and only calcium gluconate should be mixed with phosphate.

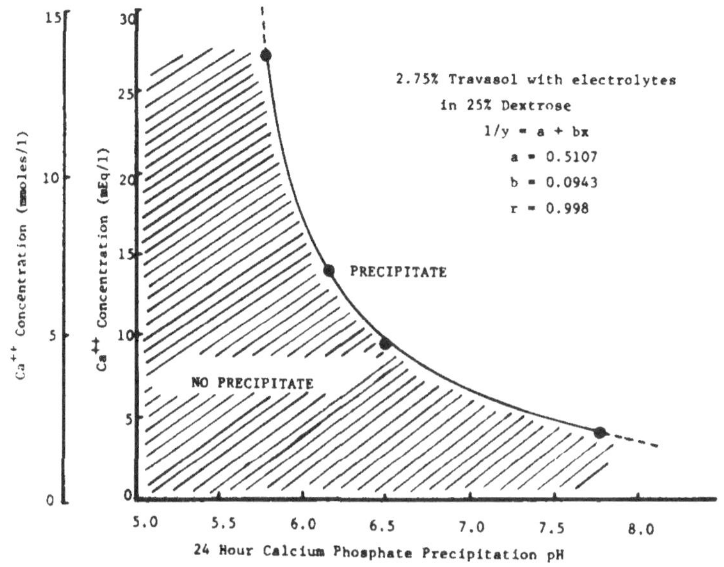

FIGURE 1. Calcium ion concentration in an extemporaneously prepared parenteral solution as a function of the admixture pH and calcium phosphate precipitation at 24 hr. (From Nedich, R. L., in *Advances in Parenteral Nutrition*, Johnston, I. D. A., Ed., University Park Press, Baltimore, 1978, 419. With permission.)

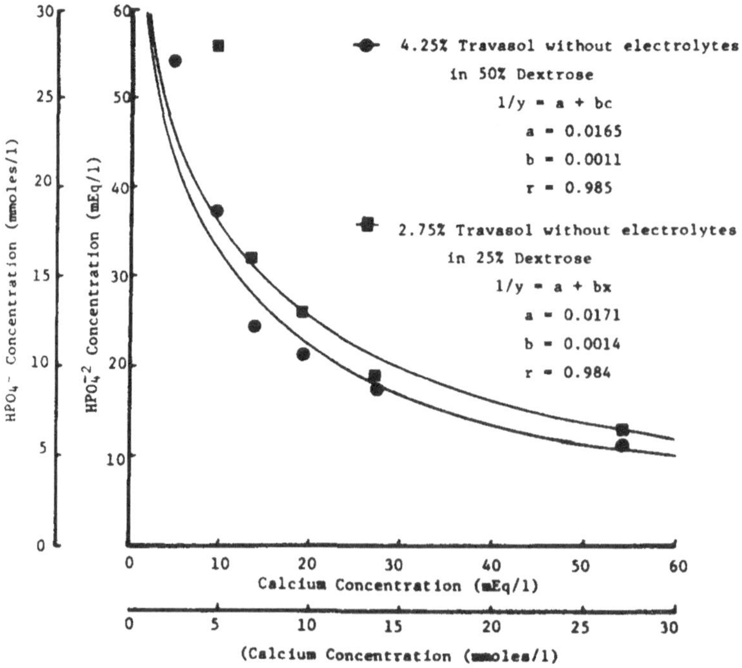

FIGURE 2. Phosphate concentration as a function of calcium concentration for calcium phosphate precipitation in parenteral nutrition mixtures at 24 hr. (From Nedich, R. L., in *Advances in Parenteral Nutrition*, Johnston, I.D.A., Ed., University Park Press, Baltimore, 1978, 419. With permission.)

B. Trace Elements

Aqueous solutions of single-entity and multiple-component trace elements manufactured by individual hospitals are quite stable, and today similar commercial products are available. Solutions of the halogen salts are recommended for optimum stability particularly in preparations containing multiple components.

Compatibility of trace elements in TPN admixtures has been investigated recently. Kartinos showed trace elements to be stable for a period of 30 days in an admixture consisting of 4.25% amino acids and 25% dextrose with electrolytes. Recent work by Pipa and Tsallas showed that the trace elements copper, iron, manganese, chromium, and iodine are stable for periods in excess of 60 days in a TPN admixture consisting of 4.25% amino acids, 25% dextrose, and other electrolytes. In contrast selenium showed a dramatic reduction in concentration (approximately 50%) following 10 days of storage in the admixture. At present this is under further investigation. Even with refrigeration, darkening in color of the TPN admixture can occur approximately 14 days following introduction of trace elements, however this was found to depend on the type of amino acid solution used. Pipa and Tsallas have found darkening of the TPN admixture in the presence of trace elements after 14 days of storage when Vamine N® (Pharmacia, Montreal, Quebec) was the source of amino acids. Until further studies are done to clarify this color change in terms of chemical stability of the admixture, addition of trace elements to this TPN solution should be done just prior to infusion.

IV. COMPATIBILITY AND STABILITY OF VITAMINS IN TPN SOLUTIONS

The physical and chemical stability of vitamins in parenteral nutrition solutions is subject to much debate at present. Trials have shown that multivitamin preparations are physically compatible with various commonly used intravenous infusion fluids, including protein hydrolysates and dextrose (see Table 1). However vitamin A has been found (Howard et al.) to adsorb to the polyvinylchloride used for the bags and tubing, with approximately a 30% reduction of the original concentration in the first 8 hr. Hartline and Zachman have also found a 50% reduction in the first 24 hr. Because of the uncertain stability of vitamins in TPN admixtures it is recommended that until further work is done in this area vitamins be added to TPN solutions just prior to infusion.

V. COMPATIBILITY AND STABILITY OF ANTIBIOTICS AND OTHER DRUG ADDITIVES TO TPN SOLUTIONS

A. Antibiotics

The addition of antibiotics to TPN admixtures should not be done unless there is no alternative. Where possible they should be given separately or through a Y tubing, or into another vein. However sometimes it becomes necessary to add antibiotics and other drugs to the TPN solution because fluid intake must be restricted or peripheral veins are not accessible. Most compatibility studies with drugs are based on visual observations such as clarity of solution, color change, precipition or pH, and not on chemical stability. Schnetz and Kurg examined the compatibility of ampicillin, cephalothin, kanamycin, and gentamycin in a TPN admixture containing 4.0% amino acids and 25% dextrose with electrolytes and vitamins. Ampicillin, kanamycin, cephalothin, and gentamycin were found physically compatible and microbiologically stable for 12 hr in the amino acid/dextrose solution. Addition of calcium and phosphate salts affected the stability of the antibiotics. Elevated levels of other electrolytes resulted in a precipitate within 12 hr. Ampicillin appeared to be easily precipitated in the presence

Table 1
PHYSICAL COMPATIBILITY OF M.V.I.® WITH VARIOUS TPN MIXTURES

M.V.I.® may be used with most parenteral admixtures and additives. Tests show that M.V.I.® is physically compatible when mixed with the following

- Achromycin® 500 mg[a]
- ACTH 40 units
- Adrenalin® 4 mg[a]
- Alcohol 5% and dextrose 5% in water
- Aldomet®
- Aminophylline 500 mg
- Aminosol 5% with dextrose 5% solution
- Aminosyn®
- Aureomycin® 500 mg[a]
- Caffeine and sodium benzoate 500 mg[a]
- Chloromycetin®
- Sodium succinate 1 g[a]
- Decadron® phosphate 20 mg[a]
- Dextran 6% in saline
- Dextrose 5% in saline
- Dextrose 10% in saline
- Dextrose 2.5% in half-strength saline
- Dextrose 5% in half-strength saline
- Dextrose 10% in half-strength saline
- Dextrose 5% in water
- Dextrose 10% in water
- Diamox® 375 mg[a]
- Diuril® 2 g[a]
- Ephedrine sulfate 50 mg[a]
- FreAmine® II
- Ionosol MB with dextrose 5%
- Ionosol G with dextrose 10%
- Ionosol T with dextrose 5%
- Ionosol D with dextrose 10%
- Lactated Ringers injection
- Lincomycin 600 mg[a]
- Oxytetracycline 500 mg[a]
- Penicillin G 2 million units
- Potassium chloride
- Potassium phosphate
- Sodium chloride injection
- Sodium Lactate 1/6 molar

[a] Drug mixed with 5% dextrose in water.

Data from M.V.I. -1000 Monograph, Technical Data, USV Laboratories, Tuckahoe, N.Y., 1979. With permission.

of an electrolyte. By contrast cephalothin was resistant to precipitation. However, cephalothin is incompatible with calcium gluconate. Stabilities of kanamycin, cephalothin, and gentamycin were markedly decreased in an amino acid/dextrose/electrolyte TPN admixture supplemented with vitamins, whereas that of ampicillin increased under the same circumstances.

More recent work by Farago (Table 2) has shown that cefazolin, clindamycin, cloxacillin, gentamycin, and penicillin G potassium are physically compatible and microbiologically stable for 24 hr at 22°C in a TPN solution consisting of 4.25% amino acids and 16.65% dextrose with electrolytes and trace elements. Furthermore, gentamycin was both physically and microbiologically stable for 24 hr at 22°C when combined with cefazolin, clindamycin, or penicillin G potassium in the same TPN solution. In these studies cloxacillin appeared to be compatible with gentamycin. Lynn et al., however, have reported incompatibility when cloxacillin and gentamycin were admixed in an aqueous solution. Therefore until further studies are done to verify the compatibility of these two antibiotics it is recommended that cloxacillin and gentamycin not be injected into the same TPN solution.

Addition of amphotericin to TPN solutions has resulted in precipitation of the amphotericin.

Using the above principle, the Toronto General Hospital has given antibiotics in the TPN admixture in the treatment of infections in patients on home TPN with excellent clinical results and bacteriologically confirmed eradication of the infection.

B. Compatibility of Insulin

In a number of institutions the addition of exogenous insulin to TPN solution has become a common practice. Although no physical incompatibility has been demon-

Table 2
COMPATIBILITY OF ANTIBIOTICS WITH TPN-SOLUTIONS[a]

Antibiotic	Compatibility[b]	Temperature (°C)	Time (hr)	Initial conc or % loss	Solution[c]	Ref.
Ampicillin	C	4	24	1 g/ℓ	1	Feigin et al.
	C	—	6	20 g/ℓ	2	Pharmacia
	C	22	12	—	3	Pharmacia
	IC-LA	25	6	10%	1	Feigin et al.
			12	15%		
			24	25%		
	IC-PM	5	24	—	5	Athanikar et al.
Carbenicillin	C	4	24	1 g/ℓ	1	Feigin et al.
	C-PM	4	24	8 g/ℓ	5	Athanikar et al.
		25	24	—	1	Feigin et al.
Cefazolin	C	22	24	1 g/dℓ	4	Farago
	C-PM	4	24	1 g/ℓ	5	Athanikar et al.
Cephalothin	C	25	24	1 g/ℓ	1	Feigin et al.
	C	22	12	—	3	Schuetz et al.
	C-PM	4	24	2 g/ℓ	5	Athanikar et al.
	IC-LA	25	24	13%	1	Feigin et al.
	IC-PT	22	8	—	3	Schuetz et al.
Clindamycin	C	37	24	250 mg/ℓ	1	Feigin et al.
		25				
		4				
	C	22	24	300 mg/dℓ	5	Farago
	C-PM	4	24	600 mg/ℓ	5	Athanikar et al.
Cloxacillin	C	—	6	20 g/ℓ	2	Pharmacia
	C	22	24	1 g/dℓ	4	Farago
Erythromycin gluceptate	C-PM	4	24	1 g/ℓ	5	Athanikar et al.
Gentamicin	C	22	12		3	Schuetz et al.
	C	22	24	80 mg/dℓ	4	Farago
	IC-PT	22	8—24	80 mg/ℓ	3	Schuetz et al.
Kanamycin	C	4	24	250 mg/ℓ	1	Feigin et al.
	C	22	12	—	3	Schuetz et al.
	C-PM	4	24	500 mg/ℓ	5	Athanikar et al.
	IC-LA	25	6	12%	1	Feigin et al.
			12	18%		
			24	39%		
Lincomycin	C	—	12	1.2 g/ℓ	2	Pharmacia
Methicillin	C	25	24	250 mg/ℓ	1	Feigin et al.
	C	—	6	20 g/ℓ	2	Pharmacia
	C-PM	4	24	1 g/ℓ	5	Athanikar et al.
	IC-LA	25	24	20%	1	Feigin et al.
Oxacillin	C-PM	4	24	500 mg/ℓ	5	Athanikar et al.
Penicillin G potassium	C	25	24	5 M units/ℓ	1	Feigin et al.
	C	—	6	60 g/ℓ	2	Pharmacia
	C	22	24	2.5 M units/ 100 mℓ	4	Farago
	C-PM	4	24	1 M units/ℓ	5	Athanikar et al.
	IC-LA	25	24	—	1	Feigin et al.
Sulfonamides	IC-PT[d]	—	—	—	—	Dryps and Hoffman
Tetracyclines	C-PM[e]	4	24	500 mg/ℓ	5	Athanikar et al.
	IC-PT	—	—	—	2	Pharmacia
Tobramycin	C-PM	4	24	80 mg/ℓ	5	Athanikar et al.

Note: For explanation of footnotes see Table 3.

strated with TPN admixtures, there is concern that this hormone is adsorbed to the surfaces of various tubings and containers, both glass and plastic. However, the literature appears to be confusing and contradictory. Some investigators have found a 10 to 12% loss while others state that as much as 55% may be adsorbed to the tubing. Work done by Eli Lilly and Travenol Laboratories has shown that 80 to 90% of the insulin added to TPN solutions is available for 24 hr. Furthermore, clinically, the addition of insulin to TPN solutions appears to control glucose levels effectively.

C. Heparin

Heparin is added not only to TPN solution but also to the fat emulsion. Heparin has been found physically stable in both.

D. Other Drugs

Addition of pressor agents, sulfonamides, and digitalis should be avoided. Although it is commonly recommended that steroids not be added to TPN solution, hydrocortisone and methyl prednisolone have been added and found physically compatible.

In general, because of the lack of conclusive data the addition of drugs to TPN solutions should be avoided, but if certain drugs must be given with the TPN solution the pharmacy should be consulted regarding the stability and compatibility of these agents. Table 3 provides a comprehensive list of drug additives found physically and chemically compatible with TPN solutions.

VI. STABILITY AND COMPATIBILITY OF FAT EMULSIONS

Fat emulsions available in North America contain either soybean oil or safflower oil and are stable at room temperature over periods exceeding 18 months. Various in vivo studies performed with the soybean oil fat emulsion indicate that despite conventional teaching, antibiotics, electrolytes, vitamins, and amino acid solutions may be added to fat emulsions. These results are not conclusive enough to be extrapolated to clinical practice. Therefore, until more definitive studies become available the injection of any additives, with the exception of heparin, 1 unit/mℓ of emulsion, as well as the admixing of fat emulsion with amino acid-dextrose solution in the same container, is not recommended. Furthermore, fat emulsions should be kept refrigerated unless otherwise specified by the manufacturer. Studies have demonstrated that fat emulsions support bacterial growth readily and thus as a precaution they should not be kept longer than 24 hr at room temperature.

VII. COMPATIBILITY OF TPN ADMIXTURES WITH TYPE OF CONTAINER

All TPN admixtures of amino acid-dextrose with electrolytes are fairly stable in both glass and plastic containers. One should be aware, however, of the adsorption of various additives to plastic containers (e.g., PVC). Furthermore, surfactants, or fat-soluble substances acting like surfactants in the admixture, can cause leaching of plasticizers from tubing and containers. Studies have shown that in 48 hr approximately 2.5 mg of DEHP can be leached out by 200 mℓ of a 0.04% solution of Polysorbate® 80. Aqueous solutions stored in PVC bags show low levels of the principle plasticizer di-(2-ethylhexyl) phthalate (DEHP) used in the polymerization of the plastic. The concentration of this agent does not exceed 0.05 ppm at room temperature when observed up to the expiry date of the product. Rubin reported DEHP levels in aqueous protein hydrolysates, stored for more than one year under a wide range of ambient temperature, as not exceeding the analytical blank. Nedich has found DEHP levels in the range

Table 3
COMPATIBILITY OF DRUGS (OTHER THAN ANTIBIOTICS) WITH TPN SOLUTIONS[a]

Drug	Compatibility[b]	Concentration (amount/l)	Time (hr/25°C)	Solution[c]	Ref.
Ascorbic acid	C	1 g	12	2	Pharmacia
Chlorpromazine	C	200 mg	12	2	Pharmacia
Cimetidine	C	300—600 mg	24	5	Tsallas
Desacetyllana-toside C	C	0.8 mg	12	2	Pharmacia
Digoxin	IC	—	—	6,7	Burke
Dipyridamole	C	30—40 mg/l	12	5	Tsallas
Heparin	C	50,000 IU	12	2	Pharmacia
	C	500—1,000 IU	—	5	Tsallas
	C	1,000 IU	L	6	Tsallas
Hydralazine	C	50 mg	12	2	Pharmacia
Hydrocortisone acetate	C	200 mg	12	2	Pharmacia
Hydrocortisone Sodium	C	200 mg	12	2	Pharmacia
Succinate[f]	C	—	48	5	Schuetz et al.
Insulin (regular)[e]	C	100 IU	12	2	Pharmacia
	AD (47%)	30 IU	12	7	Weber
	AD	—	—	7	Tsallas
	AD (23—27%)	—	—	7	Burke
	AD (10—20%)	—	24	7	Nedich
Lidocaine	C	200 mg	12	2	Pharmacia
Melphalan	C	100 mg	12	2	Pharmacia
Metaraminol bitartrate	C	200 mg	12	2	Pharmacia
	C	—	48	5	Tsallas
Methohexital	AD	—	—	7	Moorehatch and Chiou
Metoclopramide	C	10—20 mg	24	7	Tsallas
Neostigmine	C	12.5 mg	12	2	Pharmacia
Noradrenaline	C	2 mg	12	2	Pharmacia
Oxytocin	C	10 IU	12	2	Pharmacia
Phenobarbital	C	15 mg	—	7	Tsallas
Prednisolone Succinate Sodium	C	50 mg	12	2	Pharmacia
Sodium bicarbonate[g]	IC	—	—	7	Burke
Theophylline	C	1 g	12	2	Pharmacia
Warfarin	AD	—	—	7	Moorehatch and Chiou

[a] Taken from Farago (1980), see Reading List for this chapter.
[b] Code for compatibility:

C —	compatible
C-PM —	compatible; no increase in particulate matter
IC —	incompatible
IC-LA —	incompatible; loss of activity
IC-PM —	incompatible; increase in microscopic particles
IC-PT —	incompatible; visible precipitate formed
AD(%) —	adsorption to polyvinyl chloride bag

Table 3 (continued)
COMPATIBILITY OF DRUGS (OTHER THAN ANTIBIOTICS) WITH TPN SOLUTIONS[a]

[c] Code of solution with which item was tested for compatibility:

(1)	Freamine® (McGaw) crystalline amino acids	1.7%
	Dextrose	20%
	Electrolytes (m*M*): K 30, Cl 40, Mg 4.0, PO_4 7.5, Na 40, Ca 4.8 , acetate 15	
	M.V.I.® - 1000[h] (Arlington Laboratories)	10 m*l*/*l*
(2)	Vamin® (Pharmacia) crystalline amino acids	7%
	Fructose	10%
	Electrolytes (m*M*): Na 50, Ca 2.5, Cl 55, K 20, Mg 1.5	
(3)	Crystalline amino acids (Cutter)	4%
	Dextrose	25%
	Electrolytes (m*M*): Na 40, Mg 3, acetate 50, K 30, Cl 50	
(4)	Travasol® (Baxter-Travenol Laboratories)	4.25%
	Dextrose	16.65%
	Electrolytes (m*M*): Na 55, Ca 3.6, Cl 100.4, K 70, Mg 4.5, $PO_4$15	
(5)	Travasol® (Baxter-Travenol Laboratories)	3%
	Dextrose	17.5%
	Electrolytes (m*M*): Na 58, Ca 3.85, Cl 71.5, K 30.9, Mg 5.8, PO_4 12	
	Trace Elements[i]	
(6)	Nutralipid® 10% lipid emulsion	
(7)	Crystalline amino acids-dextrose solution, with varying concentrations of amino acids, dextrose, and electrolytes	

[d] Precipitation of the water-insoluble acid form may occur if a sulfonamide is introduced into the tubing during administration of a TPN solution (Dryps et al.).

[e] Calcium and magnesium ions may form insoluble complexes with tetracyclines (McGaw).

[f] The incompatibility of magnesium sulfate with hydrocortisone sodium succinate has been reported to occur when mixed in the same syringe (anon.) However, no precipitation has occurred when they were added using separate syringes and the solution was well mixed after each addition.

[g] Addition may result in a loss of bicarbonate due to the evolution of carbon dioxide (Dryps et al.). The activities of insulin and vitamin B complex are decreased in the presence of sodium bicarbonate, and precipitation of calcium and magnesium may occur (Burke).

[h] M.V.I.®-1000 (10 m*l*) — vitamin C 1,000 mg, vitamin A 10,000 IU, vitamin D 1,000 IU, vitamin E 10 IU, thiamine HCl 25 mg, riboflavin 10 mg, niacinamide 100 mg, pyridoxine HCl 12 mg, D-pantothenic acid 26 mg.

[i] Trace elements (mg/*l*): Cu 1.0, Mn 0.7, Zn 3.0, I^- 0.12, Cr 0.002, Se 0.12

of 0.01 to 0.2 ppm in TPN solutions containing 2.75% amino acids with 25% dextrose and 4.25% amino acid with 25% dextrose stored in PVC bags at 5°C over a period of 30 days.

Adsorption to plastic is definitely a major problem with drugs such as barbiturates and vitamins. In particular, vitamin A has been found to be adsorbed by plastic, resulting in a loss of as much as 75% of its initial concentration within 24 hr. Hence components of a lipid nature and those with surfactant activity should not be packaged in PVC plastic containers.

Although glass containers are free of the above-mentioned problems, their use is associated with other interactions, i.e., between rubber closures of the glass container and certain additives, such as trace metals, drugs, etc.

With regard to microbial contamination the PVC units, unlike glass containers, offer the advantage of a closed system requiring no air replacement in order to empty, thus minimizing the chances of contamination.

VIII. SUMMARY ANNOTATED LIST OF DRUGS AND OTHER ADDITIVES COMPATIBLE WITH TPN SOLUTIONS*

A. Electrolytes

•Sodium and Potassium	— usually the chloride or acetate salt is compatible, quantities quite variable
•Calcium	— as calcium gluconate (much more compatible than the chloride salt) — can interact with phosphate to form insoluble calcium phosphate precipitate — interaction is temperature-, pH-, and concentration-dependent — interaction with phosphate shown to be more rapid when calcium is used as the chloride
•Magnesium	— usually as the sulfate or chloride salt — sulfate salt may interact with calcium to form insoluble calcium sulfate, a reaction that is concentration dependent
•Phosphate	— usually as the potassium salt — to be added with caution when a calcium salt is present, and prior to the addition of the calcium
•Bicarbonate	— usually as the sodium salt — caution is needed when adding to solution containing calcium, insoluble calcium carbonate may form — given only if necessary and added just prior to TPN infusion — decreases activity of insulin and vitamin B-complex with vitamin C — acetate or lactate salts are preferable to bicarbonate

B. Trace Elements

•Copper	— 1 mg/ℓ as cupric chloride
•Zinc	— 3 to 15 mg/ℓ as zinc sulfate or the chloride salt
•Chromium	— 2 to 250 μg/ℓ as the nitrate or chloride salt
•Selenium	— 120 μg/ℓ as selenious acid or selenomethionine
•Iodine	— 120 μg/ℓ as the potassium or sodium iodide salt
•Manganese	— 0.7 mg/ℓ as manganous chloride
•Iron	— as iron dextran, 0.5 to 100 mg of Fe/ℓ

C. Drugs

Added just prior to administration of TPN solution:

•Insulin	— dose varies — adsorption to polyvinyl chloride bag — must be added just prior to administration to minimize adsorption

* For greater detail, consult Tables 1 to 3.

•Heparin	— 500 to 1000 units/ℓ — in special cases as high as 25,000 units has been used — 500 units added per 500 mℓ of Intralipid® 10%
•Dipyridamole (Persantine)®	— 30 to 40 mg/ℓ — gives a fluorescent appearance to the solution — used only in special patients with blood clotting problems — found physically stable up to 12 hr
•Cimetidine (Tagamet®)	— 300 to 600 mg/ℓ — found chemically stable up to 24 hr at room temperature
•Metoclopramide (Maxeran®)	— physically and chemically stable for 24 hr in doses of 10 to 20 mg/ℓ
•Phenobarbital	— physically stable at 15 mg/ℓ

D. Vitamins

Added just prior to TPN infusion:

•M.V.I.®-1000	— 5 to 10 ml/ℓ — physically compatible — (5 mℓ contains: vitamin A, 5000 IU; vitamin B_1, 22.5 mg; vitamin B_2, 5 mg; niacinamide, 50 mg; pantothenic acid, 13 mg; vitamin B_6, 6 mg; vitamin C, 500 mg; vitamin D, 500 IU; and vitamin E, 5 IU)
•Soluzyme® with Ascorbic Acid	— 1 vial (about 10 mℓ) per liter — 1 vial contains: vitamin B_1, 10 mg; vitamin B_2, 10 mg; niacinamide, 250 mg; pantothenic acid, 45 mg; vitamin B_6, 5 mg; vitamin B_{12}, 25 μg; vitamin C, 500 mg; and folic acid, 5 mg — replaces folic acid and vitamin B_{12}.
•Folic acid	— 10 to 15 mg/ℓ — not always necessary if Soluzyme® with Ascorbic Acid is added
•Vitamin B_{12}	— 1000 μg/ℓ — not always necessary if Soluzyme® with Ascorbic Acid is used
Vitamin K (Synkavite®)	— 10 to 15 mg/ℓ

E. Antibiotics Found to be Physically and Microbiologically Stable in TPN Admixtures

Cefazolin	Cloxacillin	Methicillin sodium
Cephalothin	Gentamicin	Carbenicillin
Clindamycin	Pencillin G potassium	Kanamycin sulfate

IX. MICROBIOLOGICAL STABILITY OF TPN SOLUTIONS

Microbiological growth is a major complication with TPN admixtures. Although these solutions are prepared under controlled aseptic conditions *Candida septicemia* does occur from time to time. As a result of this problem, a number of investigators have examined the nature of bacterial growth in TPN solutions. Results of various experimental studies have shown that *C. albicans* is the only microorganism that grows significantly in various TPN solutions stored at room temperature. In particular, protein hydrolysates supported a prolific growth of candida when compared with crystalline amino acids or amino acids and dextrose solutions. Fat emulsions also support bacterial growth including *C. albicans* when stored at 25 and 37°C. Dilute dextrose-containing TPN solutions will not support the growth of candida and concentrated dextrose-containing TPN solutions will destroy microorganisms. Despite these observations, the probability of microbial growth and survival in TPN solutions must be seriously considered at all times.

To avoid infection and bacterial contamination, solutions should be used within 24 hr of being brought to room temperature and should otherwise be kept at 4°C until administered.

READING LIST

Anon., Incompatibility of magnesium sulphate and hydrocortisone sodium succinate, *Am. J. Hosp. Pharm.*, 35, 783, 1978.

Athanikar, N., Boyer, B., Deamer, R., Harrison, H., Henry, R. S., Mett, C., Sturgeon, R., van Leuven, M., and Welco, A., Visual compatibilities of thirty additives with a parenteral nutrition solution, *Am. J. Hosp. Pharm.*, 36, 511, 1979.

Bergman, H. D., Incompatibilities in large volume parenterals, *Drug Intell. Clin. Pharm.*, 11, 345, 1977.

Burke, W. A., Preparation, including incompatibilities and instability, in *Symposium on Total Parenteral Nutrition*, American Medical Association, Chicago, 1972, 175.

Chichester, C. O., Stadtman, F. H., and Mackinney, G., On the products of the Maillard reaction, *J. Am. Chem. Soc.*, 73, 3418, 1951.

Cluxton, R. J., Jr., Some complexities of making compatibility studies in hyperalimentation solutions, *Drug Intell. Clin. Pharm.*, 5, 177, 1971.

Deitel, M. and Kaminski, V. M., Growth of common bacteria and *Candida albicans* in 10% soybean oil emulsion, *Can. J. Surg.*, 18, 531, 1975.

Dryps, J. S. and Hoffman, R. P., Hyperalimentation review, *Drug Intell. Clin. Pharm.*, 7, 413, 1973.

Farago, S. J., Practical aspects of total parenteral nutrition solution preparation: Studies of Antibiotic Compatibility, Stability and Sterility, Hospital Pharmacy Residency project, Women's College Hospital, Toronto, 1980.

Feigin, R. D., Okamoto, G. A., and Wyatt, R. G., Stability of antibiotics in parenteral solutions, *Pediatrics*, 49, 22, 1972.

Feigin, R. D., Moss, K. S., and Shackelford, P. G., Antibiotic stability in solutions used in intravenous nutrition and fluid therapy, *Pediatrics*, 51, 1016, 1973.

Frank, J. T., Intralipid compatibility study, *Drug Intell. Clin. Pharm.*, 7, 351, 1973.

Haugaard, G., Tumerman, L., and Silvestri, H., A study on the reaction of aldoses and amino acids, *J. Am. Chem. Soc.*, 73, 4594, 1951.

Hartline, J. V. and Zachman, R. D., Vitamin A delivery in total parenteral nutrition solutions, *Paediatrics*, 58, 448, 1976.

Hays, D. P. and Mehl, B., I. V. drug incompatibilities — insulin, *Am. J. I.V. Therapy*, April/May, 30, 1976.

Henry, R. S., Jurgens, R. W., Jr., Sturgeon, R., Athanikar, N., Welco, A., and Van Leuven, M., Compatibility of calcium chloride and calcium gluconate with sodium phosphate in mixed T.P.N. solutions, *Am. J. Hosp. Pharm.*, 37, 673, 1980.

Hirsh, J. I., Fratkin, M. J., Wood, J. H., and Thomas, R. B., Clinical significance of insulin adsorption by polyvinyl chloride infusion systems, *Am. J. Hosp. Pharm.*, 34, 583, 1977.

Howard, L., Chu, R., Feman, S., Mintz, H., Ovesen, L., and Wolf, B., Vitamin A deficiency from long-term parenteral nutrition, *Ann. Intern. Med.*, 93, 576, 1980.

Hull, R. L., Physiochemical considerations in intravenous hyperalimentation, *Am. J. Hosp. Pharm.*, 31, 236, 1974.

Istchenko, M. and Tsallas, G., An Experiment to Determine Calcium Intolerance with Magnesium and Phosphate in T.P.N. Solutions Manufactured By Toronto General Hospital, Canada, 1976.

Johnson, C., Cloyd, J., and Kapp, R. P., Parenteral hyperalimentation, *Drug Intell. Clin. Pharm.*, 9, 496, 1975.

Kaminski, M. V., Jr., Harris, D. F., Collin, C. F., and Sommers, G. A., Electrolyte compatibility in a synthetic amino acid hyperalimentation solution, *Am. J. Hosp. Pharm.*, 31, 244, 1974.

Kartinos, N. J., Trace elements formulations in intravenous feeding, in *Advances in Parenteral Nutrition*, Johnston, I.D.A., Ed., M.T.P. Press, Lancaster, England, 1978, 233.

Kleinman, L. M., Tangrea, J. A., Gallelli, J. F., Brown, J. H., and Gross, E., Stability of solutions of essential amino acids, *Am. J. Hosp. Pharm.*, 30, 1054, 1973.

Klotz, R., Sherman, J. O., and Egan, T., Preparation of hyperalimentation solutions for the pediatric patient, *Am. J. Hosp. Pharm.*, 28, 102, 1971.

Kobayashi, N. H. and King, J. C., Compatibility of common additives in protein hydrolysate/dextrose solutions, *Am. J. Hosp. Pharm.*, 34, 589, 1977.

Kramer, W., Inglott, A., and Cluxton, R., Some physical and chemical incompatibilities of drugs for i.v. administration, *Drug Intell. Clin. Pharm.*, 5, 211, 1971.

Laegeler, W. L., Til, J. M., and Blake, M. I., Stability of certain amino acids in a parenteral nutrition solution, *Am. J. Hosp. Pharm.*, 31, 776, 1974.

Lynn, B., Pharmaceutical aspects of semi-synthetic penicillins, *J. Hosp. Pharm.*, 28, 71, 1970.

Lynn, B. and Jones, A., Advances in Antimicrobial and Antineoplastic Chemotherapy 1, *Urban and Schwarzenberg*, Munich, 1972, 701; quoted in *Martindale: The Extra Pharmacopoeia*, 27th ed., Wade, A., Ed., Pharmaceutical Press, London, 1977, 119.

Mada, P. L., Madan, D. K., and Palumbo, J. F., Total parenteral nutrition, *Drug Intell. Clin. Pharm.*, 10, 693, 1976.

Moorehatch, P. and Chiou, W. L., Interactions between drugs and plastic intravenous fluid bags, *Am. J. Hosp. Pharm.*, 31, 72, 1974.

Nedich, R. L., The compatibility of extemporaneously added drug additives with Travasol® (amino acid) injection, in *Advances in Parenteral Nutrition*, Johnston, I. D. A., Ed., M.T.P. Press, Lancaster, England, 1978, 415.

Product Monograph, Parenteral Nutrition: Vamin® and Nutralipid®, Pharmacia (Canada) Ltd., Dorval, Quebec, 1978, 38.

Parker, E. A., Compatibility digest, *Am. J. Hosp. Pharm.*, 27, 492, 1970.

Patrick, R. J., Loucas, S. P., Cohl, J. K., and Mehl, B., Review of current knowledge of plastic intravenous fluid containers, *Am. J. Hosp. Pharm.*, 34, 357, 1977.

Pinkus, T. F. and Jeffrey, L. P., Incompatibility of calcium and phosphate in parenteral alimentation solutions, *Am. J. I. V. Therapy*, Feb./Mar., 22, 1976.

Product Monograph, Freamine®, McGaw Co., Glendale, Calif., 1971.

Rowlands, D. A., Wilkinson, W. R., and Yoshimura, N., Storage stability of mixed hyperalimentation solutions, *Am. J. Hosp. Pharm.*, 30, 436, 1973.

Rubin, R., Storage of aqueous solutions for parenteral infusion, *Lancet*, 1, 965, 1972.

Schroeder, L. J., Iacobellis, M., and Smith, H. A., The influence of water and pH on the reaction between amino compounds and carbohydrates, *J. Biol. Chem.*, 212, 973, 1955.

Schuetz, D. H. and King, J. C., Compatibility and stability of electrolytes, vitamins, and antibiotics in combination with 8% amino acids solution, *Am. J. Hosp. Pharm.*, 35, 33, 1978.

Tsallas, G., Drugs and other additives added to total parenteral nutrition solutions prepared at Toronto General Hospital, *Nutritional Support 1979: Management of Malnutrition in Hospitalized Patients*, University of Toronto, Canada, 1979, 28.

Walfrom, M. L., Kolb, D. K., and Langer, A. W., Jr., Chemical interactions of amino compounds and sugars. VII. pH dependency, *J. Am. Chem. Soc.*, 75, 3471, 1953.

Weber, S. S., Wood, W. A., and Jackson, E. A., Availability of insulin from parenteral nutrient solutions, *Am. J. Hosp. Pharm.*, 34, 353, 1977.

Weisenfeld, S., Podolsky, S., Lowell, G., and Lawrence, Z., Adsorption of insulin to infusion bottles and tubing, *Diabetes*, 17, 766, 1968.

West, K. R., Sansom, L. N., Cosh, D. G., and Thomas, M. P., Some aspects of the stability of parenteral nutrition solutions, *Pharm. Acta Helv.*, 51, 19, 1976.

Wilkinson, W. R., Flores, L. L., and Pagones, J. N., Growth of microorganisms in parenteral nutritional fluids, *Drug Intell. Clin. Pharm.*, 7, 226, 1973.

Chapter 18*

HOME CARE, TPN, AND THE PHARMACIST

George Tsallas

The availability of TPN therapy capable of being given at home has opened a new and challenging area for the pharmacist. The extent and type of pharmaceutical service has been outlined previously under the pharmacist's role as a member of the TPN team. While recognizing that the type and scope of the service of the pharmacy will vary in different ambulatory programs, it is nevertheless possible to give the following general guidelines and options.

By and large, the pharmacy has become responsible for the provision of nutrient solutions including related supplies such as syringes, needles, etc. However, the mechanics of this service varies significantly among institutions. The pharmacy in some institutions offering this service, particularly in the U.S., trains their patients to make their own TPN admixtures at home. After discharge of the patient the pharmacy becomes responsible for the provision of commercial stock solutions to these patients to serve as the raw materials for making their own nutrient infusates. By contrast other institutions including the Toronto General Hospital provide the finished TPN admixture and related supplies at monthly intervals.

At the Toronto General Hospital the Pharmacy department schedules and prepares monthly the entire needs of the patients. In addition the pharmacy will procure and administer all related supplies required by the patient. A Patient TPN Requisition (Figure 1) is filled by the patient and mailed to the pharmacy once a month. A pharmacy technician will fill the order, then have a pharmacist check it prior to releasing the material to the patient.

Although this centralized system of preparing the solutions may not have the cost advantage of a system that requires the preparation of such solutions by the patient, it has several advantages. These include:

1. Time for the patient to pursue activities other than that of preparing the solutions — a time-consuming affair.
2. Help to patients with limited space, such as those residing in a small apartment.
3. Reduction of visits to pharmacy due to the pharmacist's discussion with the patient of his "shopping list" (Figure 1) resulting in fewer oversights.
4. More rapid changes of the formula when necessary without having to wait for re-education of the patient.
5. More stringent quality control, including the aspects of composition and contamination.
6. More efficient reaction to emergency situations because of the superior facilities of the pharmacy and the usually better communication between physician and pharmacist than between physician and patient.
7. Closer contact with the patient and an appreciation of his/her daily routine when at home, thus contributing to better morale for both patient and pharmacist.

More recently in certain parts of North America, and in particular in the U.S., provision of the home-care daily needs (including solutions, training, cost reimbursement, and monitoring) is being offered by commercial enterprises. These organizations are staffed with pharmacists and nurses to provide for the total needs of the patient, in-

* Reading List follows Chapter 19.

PATIENT'S NAME: ____________________

379[a]	DRESSINGS AND BANDAGES	UNIT SIZE	NUMBER REQUESTED	NUMBER FILLED	FILLED BY[b]	CHECKED BY[c]	COST
15710	Adhesive Tape 1"	10 yds					
15579	Alcohol Swabs	150/box					
13296	Disposable Masks	50/box					
12776	Disposable Sterile Gloves	100/box					
15778	Micropore Tape 2"	10 yds					
15777	Non-allergenic Micropore Tape 1"	10 yds					
521914	Non-allergenic Waterproof Tape 2"	5 yds					
St.Rm.	Op-Site(R) (small)	5x7.5cm					
10275	Op-Site(R) (large)	10x14cm					
11860	Sterile Cotton Balls	25/box					
C.S.S.	Sterile Gauze ½" x ½"	1/pkg					
14711	Sterile Gauze 2" x 2"	50/box					
14726	Sterile Gauze 4" x 4"	50/box					
380	NEEDLES AND SYRINGES						
13530	Disposable Needles 18 G 1½"	100/box					
13515	Disposable Needles 21 G 1½"	100/box					
13550	Disposable Needles 22 G 1½"	100/box					
13525	Disposable Needles 25 G 5/8"	100/box					
961946	Disposable Syringes 1 ml	100/box					
961951	Disposable Syringes 3 ml	50/box					
961945	Disposable Syringes 6 ml	50/box					
961944	Disposable Syringes 12 ml	50/box					
961942	Disposable Syringes 20 ml	20/box					
384[a]	TUBING						
521910	Medication Injection Caps	50/box					
521905	Medication Injection Sites	48/box					
St.Rm.	Non-vented IV Set (with 0.22 µ Ivex-2(R) Filter)	48/box					
13104	Nutralipid(R) Tubing	10/pkg					
521906	Plasma Transfer Set	2/box					
Aut.St.	Primary Extension Tubing 7"	1/box					
13075	Primary Extension Tubing 30"	1/box					
St.Rm.	Solution Transfer Set (with flash-ball)	1/box					
778	MISCELLANEOUS SUPPLIES						
10105	Bardic(R) Bile Bags	12/box					
16560	Complete Urine Drainage Bed Set	12/box					
-	Compressed Air Cylinder (size E)	1					
10125	Cutter(R) Urine Leg Bag	12/box					
	IV SOLUTIONS						
TGH	Amino Acid Dextrose Solution (added electrolytes)						
374200	Dextrose/Water 5%	250 ml					
003200	Nutralipid(R) 10%	500 ml					
891302	Sodium Chloride (with bacteriostat)	30 ml					
C.S.S.	Sodium Chloride	250 ml					
C.S.S.	Sodium Chloride	500 ml					
C.S.S.	Sodium Chloride	1000 ml					
153[a]	DRUGS AND MEDICINALS						
116505	Detergicide 1:750	500 ml					
St.Rm.	Dextrostix(R)	25/box					
St.Rm.	Diastix(R)	50/box					
211301	Heparin (1000 units/ml)	10 ml					
TGH	Iron Dextran (0.5 mg/ml)	10 ml					
736526	Isopropanol 80%	500 ml					
361903	Ketostix(R)	50/box					
St.Rm.	Povidone-Iodine Ointment 10%	1 gm					
731519	Povidone-Iodine Solution 10%	500 ml					
131201	Soluzyme(R)	10 ml					
144201	Synkavite(R) (10 mg/ml)	1 ml					
TGH	Vitamin A Injection (2500 iu/ml)	10 ml					
	OTHERS						

FOR PATIENT USE: Date Order Sent: ____________________

Date To Be Picked Up: ____________________

Comments:

FOR OFFICE USE ONLY: Date Order Filled: ____________________ By: ____________________ (Technician's Signature)

Comments:

Footnotes:

a) Code for hospital administrative costing.
b) Technician.
c) Pharmacist.

FIGURE 1. Replica of pharmacy (monthly) Order Sheet for home TPN patients.

cluding delivery of solutions and regular reporting to the physician as well as to the pharmacist and nursing personnel of the responsible institution. This service seems to have tremendous potential for the future in home TPN, particularly for institutions wanting to provide such services without having the requisite pharmacy staff and for institutions with such services but where the patient is at great distance, preventing readily accessible supply of mixtures.

Chapter 19

COST ANALYSIS FOR HOME CARE TPN

George Tsallas

Total parenteral nutrition (TPN) is an expensive form of therapy. In-hospital patient costs may be covered by government support or by third-party insurance. Home care therapy however is most likely to become a financial burden for the patient, particularly since private insurance companies usually resist covering such costs. Thus the majority of patients requiring home care TPN will not be able to meet the cost of therapy without financial assistance.

In certain parts of North America, such as Canada, financial assistance is sought by individual hospitals capable of providing this service through the federal and/or provincial governments.

The major costs of home care TPN services include the cost of solutions, infusion equipment, dressing supplies, and certain equipment for the preparation, storage, and administration of solutions. Various reports in the literature show that costs vary according to the composition of the solutions and the type of equipment used for the infusion of the nutrient solutions. Such reports show the cost to range from $18 to $40 per patient per day, but nevertheless with savings of up to 60% by comparison with the cost of treatment of these patients in the hospital.

The home care TPN program at the Toronto General Hospital (TGH) is financed by the (provincial) Government of Ontario Ministry of Health and is classified as a life support program.

The cost accounting of home care TPN at TGH is the responsibility of the Pharmaceutical Services. Such cost accounting is prepared once every 28 days and covers the cost of solutions, tubing, filters, etc., as shown in Table 1.

As shown in Table 1, the average per diem cost per patient is $62.06 (Can.) This cost excludes pharmacy salaries and/or any special laboratory tests or procedures. By comparison, the average per diem cost per patient for similar supplies when used in hospital is $73.22 (Can.). This difference of cost between home care and inpatient therapy is obviously of little importance. However a significant difference in cost for inpatient and home care therapies is evident when the true inpatient cost takes into consideration the standard per diem private hospital rate of $285.00. Upon comparison one then finds the inpatient daily cost to be approximately four and a half times that of home care. This significant saving of 78% demonstrates well the cost-benefits of home care TPN.

For home TPN there is an additional capital expense that is basically a one-time type of cost, following the release of the patient from the hospital. This outlay is for certain equipment required for the sterilization of utensils, dressings, storage of nutrient solutions, and administration of the infusate (see Table 2).

In summary, home care TPN is costly for all parties concerned, but more so for the patient who must have outside financial assistance to afford this vital therapy. By comparison however to providing this therapy in hospital, home care TPN is recognized to be the most cost-effective way of offering this medically recognized life support service.

Table 1
AVERAGE DAILY PHARMACEUTICAL COSTS OF TPN FOR HOME CARE THERAPY AND IN-HOSPITAL PATIENTS

Items	Cost (Can.$) Home care	Inpatient
Dressings and bandages	0.58	0.17[a]
Syringes and needles	0.83	2.00[a]
Tubing	7.90	5.66
Filters	1.30	0.67
Containers	6.70	7.01
Drugs, medicinals, and i.v. solutions	44.60	57.71
Miscellaneous	0.15	—
Total cost:	62.06	73.22

[a] Estimated cost based on daily usage of these items.

Table 2
EQUIPMENT REQUIRED FOR A PATIENT DISCHARGED TO THE HOME CARE TPN PROGRAM

Item	Quantity Cost	(Can.$)
Large roasting pan (4-*l* capacity)[a]	1	25.00
Kitchen tongs	2	6.00
Dressing tray[b]	1	60.00
Blood pressure cuffs	3	169.50
Double-stage air regulator	1	176.00
i.v. Pole on wheels	1	115.00
Compressed air cylinder, size E	1	6.20
Refrigerator (19 ft^3)[a]	1	700.00
Total:		1257.70

[a] Items purchased by patient.
[b] Contents — 1 metal tray, 2 small stainless-steel basins, 1 Kelly forcep, 2 thumb forceps, 1 medicine glass.

READING LIST

(Chapters 18 and 19)

Burke, W. A., Geld, A., and Gadman, J., Total parenteral nutrition and the ambulatory patient at home, *Am. J. I. V. Ther.*, April-May, 53, 1976.

Dudrick, S. J., Englert, D. M., Van Buren, C. T., Rowlands, B. J., and MacFadyen, B. V., New concepts of ambulatory home hyperalimentation, *J. Parenteral Enteral Nutr.*, 3, 72, 1979.

Englert, D. M. and Dudrick, S. J., Prinicples of ambulatory home hyperalimentation, *Am. J. I.V. Ther.*, August-September, 11, 1978.

Ivey, M., Riella, M., Mueller, W., and Scribner, B., Long-term parenteral nutrition in the home, *Am. J. Hosp. Pharm.*, 32, 1032, 1975.

Jeejeebhoy, K. N., Langer, B., Tsallas, G., Chu, R. C., Kuksis, A., and Anderson, G. H., Total parenteral nutrition at home: studies in patients surviving 4 months to 5 years, *Gastroenterology*, 71, 943, 1976.

Scribner, B. H., Cole, J. J., Christopher, T. G., Vizzo, J. E., Atkins, R. C., and Blagg, C. R., Long-term total parenteral nutrition, *JAMA*, 212, 457, 1970.

Tsallas, G. and Baun, D. C., Home care total parenteral alimentation, *Am. J. Hosp. Pharm.*, 29, 840, 1972.

Chapter 20

PHARMACEUTICAL LOGISTICS FOR PLANNING A TPN SERVICE

George Tsallas

TABLE OF CONTENTS

I. INTRODUCTION

While the preparation of TPN solutions offers an excellent opportunity for pharmacy involvement it does present various problems that require serious consideration prior to undertaking the ultimate task.

There are four major areas to consider when planning pharmaceutical services for the preparation and delivery of TPN solutions. These include (1) facilities for storage and manufacture of solutions, (2) equipment needs, (3) personnel needs, and (4) cost. It is advisable to visit other institutions with established programs in order to see their "set up" prior to engaging in establishing one's own program. Various systems should be evaluated and one should adopt what seems suitable and practical for the type of service one wishes to provide. Budgetary restrictions may not always allow one to have the desirable or ideal arrangement. However, the safety of the patient is of primary importance and certain basic requirements for setting up this type of service must be made available.

II. FACILITIES

The preparation of these infusions demands strict sterile manufacturing techniques which cannot be carried out satisfactorily unless suitable manufacturing facilities are available for this purpose. Adequate space must be provided for ordinary storage of both raw materials (including commercial stock solutions) and refrigerated storage of finished products. In addition there is space required for aseptic compounding, sterile handling of solutions, and for the checking, labeling, and packaging of these solutions.

There should be suitable facilities for handwashing and for cleaning equipment, with a supply of apyrogenic distilled water for rinsing containers and equipment. The area should be serviced with steam, hot and cold tap water, air and air vacuum, as well as heavy electrical wiring.

The areas for compounding and sterile handling of solutions should have filtered air and preferably be under positive air pressure. In addition, the ceilings in these areas should be made of solid nonporous material and the walls of smooth washable surfaces such as glass, enameled steel, or seamless ceramic tile. The floors should be constructed of seamless vinyl or terrazzo finish for ease of cleaning, and be fitted with floor drains. All these surfaces must be readily accessible for cleaning.

The physical layout of the TPN area will usually depend on what space is made available to the pharmacy, the equipment to be used, and the method of preparing the solutions. In most established programs the layout of the physical facility is one that includes not only the compounding and the sterile handling of the solution, but also the storage of commercial concentrates, the office facilities, and the checking, labeling, and packaging of the TPN solutions. Although this arrangement may be practical it is not the most appropriate in terms of aseptic compounding and principles. Priority should be given to the most important procedures in the preparation of solutions — the aseptic compounding and sterile handling of the solutions. Other operations such as typing of labels, labeling, checking, storage, and assembling of ingredients should not be carried out in these two critical areas. Such potentially contaminating procedures should be carried out in areas outside those for admixing and sterile filling.

III. EQUIPMENT

Space requirements will also vary with the particular method (and thus the equipment) used for the preparation of TPN solutions. Basically the laminar air-flow hood (LAFH) is the most important and expensive piece of equipment, particularly when solutions are prepared individually, usually on a daily basis. Additional equipment, however, is required when solutions are to be manufactured by the bulk method. Such equipment includes a stainless steel mixing tank, a filtration apparatus, glass and steel products, pressure tubing, a self-priming pump or pressure tank, a top-loading balance, etc. Steel carts, refrigerator, waste baskets, pH meter, office equipment, etc. are needs common to all systems. The reader should refer to Chapters 14 and 15 on techniques for the manufacture of TPN solutions for a better idea of what equipment may be needed.

IV. PERSONNEL

The number of professional and technical personnel also will vary with the type of service the institution wants to provide. Individual preparation of solutions on a daily basis requires more personnel compared with the method of bulk preparation of standardized solutions. A time study at the Toronto General Hospital has shown that the average total pharmacist and technician time spent in the preparation of a 1-*l* unit of TPN solution, with an average number of 8 additives, using the bulk method technique, is 3 min. This time is total contact time with all aspects of the preparation, including setting up equipment, packaging, and storing. By comparison, the average total pharmacist and technician time spent in the preparation of the same 1-*l* unit, using the individual bottle method, is 20.9 min. It is quite evident that the method used in preparing these solutions will affect staff needs. In addition the depth of involvement in the program will also determine staff needs. Generally speaking, to provide the TPN solution needs for 10 patients receiving, on average, 3 *l*/per patient per day, the minimum staff required is one pharmacist and one trained technician working 7 hr/day/5 days/week. Among the pharmacist's responsibilities is interpretation of the physician's orders, dosage of additives, screening of formulations for physical, chemical, and pharmacological incompatibilities, determination of drug additive stability, completing the master formula form, and determination of the feasibility of compounding the solutions requested. In addition the pharmacist maintains records regarding the supply and need for solutions for patients receiving TPN. In summary, the pharmacist is responsible for the daily operation involving professional judgment and difficult pharmaceutical techniques. The technician, under the supervision of the pharmacist, compounds solutions, makes labels, does the labeling and most technical repetitive procedures, as well as maintaining the physical plant.

V. COST ESTIMATES

In the planning of a TPN pharmaceutical service, cost may become the most critical factor for the establishment of such a program. It is highly advisable that a cost estimate be prepared covering renovations (or the building of new facilities), equipment, and additional staffing needs, to be presented to the administration for approval.

When feasible, it is obvious that the building of new facilities, or even the carrying out of renovations, should be discussed with those responsible for the structure of the institution. Quite likely this aspect of cost would become the responsibility of the Planning department of the institution and not of the Pharmacy department.

On the other hand, the preparation of cost estimates for the equipment required becomes the total responsibility of the Pharmacy department. It is advisable that various suppliers be contacted and cost quotations reviewed in consultation with the Purchasing department of the institution prior to choosing a particular supplier's product. It is wise to consult other institutions which have already had experience with such equipment. It is also important that all accessories and installation costs be considered and included in the cost estimate, as well as inflationary increases during the delay between approval and actual purchase. Thus it is reasonable to include an inflationary cost factor of approximately 10% in the estimated cost of each piece of equipment.

Tabulated below is a cost estimate of the equipment required for the bulk compounding and sterilization of TPN solutions. It must be fully recognized however, that because this tabulation is based on 1981 prices, it can be used as a guide only. Obviously the estimates will change with inflation or the reverse.

A. Equipment

Item	Cost estimate per unit (Canadian $)
1. Stainless steel mixing tank, with drain valve and electrofinish, 120-ℓ capacity	2,000.00
2. Variable-speed peristaltic pump	650.00
3. 5-Way manifold and bypass valve	130.00
4. Pressure gauge with snubber	63.00
5. Silicone tubing, ½″ I.D. and ¼″ I.D.	76.00/25 ft., 74.00/10 ft.
6. Tygon® tubing	1.00/ft.
7. 142-mm or 293-mm Filter holder	800—1,000
8. Laminar air-flow hood — 6 ft wide	4,000.00
9. Gas stove/4 burner or high power hot plate	400.00
10. 142/293 Membrane prefilters (1,2,3, and 5 μm)	2.50/each
11. 0.22-μm Sterile disposable filter unit	20.00/each
12. Stainless steel carts	200.00
13. Steel and glass graduates	600.00
14. Top-loading balance	1,000.00
15. Stainless steel pliers	20.00
16. Surgical hemostats	34.00
17. Plastic cable ties	40.00/1,000
18. Cable tie gun	60.00
19. Plastic bag heat-sealer	300.00
20. Particulate matter examination station	100.00
21. Walk-in refrigerator	6,000.00
22. Y-shaped glass tubing connectors	1 .00 each
23. Metal clips	12.00/1,000
24. Metal clip hand sealer	15.00
25. Stainless steel scissors	30.00
26. 20-ℓ stainless-steel pressure vessel with gauges and 2-stage regulator[a]	550.00
27. Nitrogen cylinder, size K, with gauges and 2-stage regulator[a]	140.00
28. 15 gal/hr Still and 150-gal storage tank	12,000.00

[a] These may be used as alternates to the variable-speed peristaltic pump.

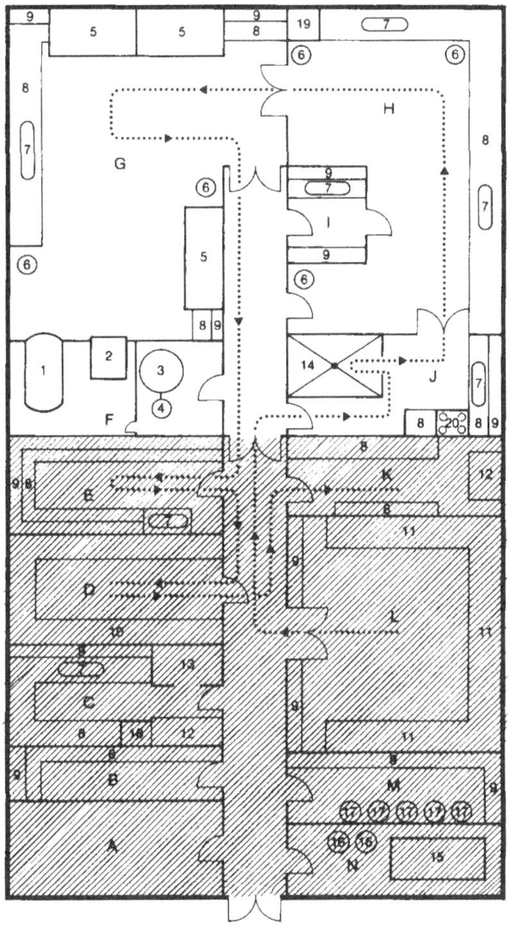

Code	Area	Sq. ft.
A	Office, clerical records	100
B	Weighing room	80
C	Quality control and research laboratory	200
D	Walk-in refrigerator	200
E	Sorting, checking, and labeling	200
F	Autoclave and still room	150
G	Sterile admixing and filling room	400
H	Sterile compounding room	300
I	Gowning and wash-up room	80
J	Equipment and commercial stock solution processing	250
K	Finished product processing for distribution	100
L	Storage of commercial stock solutions and other TPN supplies	500
M	Chemical storage room	100
N	Empty-container disposal room	100
		Total: 2760
[O]	Dirty corridor (shaded)	
[P]	Clean corridor — entire area under positive air pressure (clear)	

Code	Equipment
1	Autoclave
2	Hot-air oven sterilizer
3	150-gal distilled water storage tank
4	Glass still, 25 gal/hr capacity
5	Laminar air-flow hood
6	Disposable containers
7	Sinks
8	Work surfaces (counter tops)
9	Cupboards and shelves
10	Shelves for solution storage
11	Shelves and skids for commercial stock solutions
12	Desks
13	Fume hood
14	Equipment and bottle washer, floor drain area with shower
15	Container disposal bin
16	Broken glass disposal containers
17	Drums of chemicals
18	Benchtop refrigerator with freezer
19	Upright 20 ft^3 refrigerator
20	Gas stove or hot plates

FIGURE 1. Physical layout and area for bulk manufacture of TPN solutions.

B. Personnel

In addition to equipment and facility expenses, thought must be given to salary costs for the additional professional and technical staff needed to prepare and deliver the solutions.

VI. SUMMARY

The planning of a pharmaceutical TPN service requires that careful thought be given to the various aspects of its establishment. Of primary importance are examinations of the availability of suitable facilities and equipment, of trained personnel, and the overall cost of organizing such a service. In addition to these needs, consideration should also be given to the establishment of a TPN physician's order form, manufacturing records, a patient-monitoring form, a quality assurance program, and the establishment of a TPN committee for implementing and monitoring all TPN program activities.

Figure 1 illustrates a possible physical layout for the in-hospital preparation of TPN solutions. Modifications can be made to adapt the system to serve the particular needs of the institution concerned.

Chapter 21

UNITS OF ELECTROLYTE CONCENTRATION AND TPN FORMULATION CALCULATIONS

George Tsallas

TABLE OF CONTENTS

I. GENERAL

To facilitate explanation, most quantities in this chapter are rounded off to two or three significant figures. It is expected that in practice the user will, as a rule, consult appropriate tables for various values and carry out calculations to four significant figures in order to conclude with approximately a three-figure accuracy.

II. EQUIVALENTS (eq.) AND MILLIEQUIVALENTS (meq.)

In prescribing intravenous solutions for the treatment of disorders involving disturbances of the electrolyte balance in body fluids, the prescriber often finds it more convenient to express the concentration of such solutions in terms of milliequivalents or millimoles per liter rather than in terms of percentage or of grams per liter. An equivalent or equivalent weight is the weight in grams of an atom or a radical divided by its valence, i.e.,

$$\text{equivalent weight (eq. wt.)} = \frac{\text{wt. of atom or radical (ion) (g)}}{\text{valence of this atom or radical}}$$

The Eq. wt. is usually expressed in grams (g).

A milliequivalent (meq) or milliequivalent weight (meq. wt.) of an atom or radical (ion) is a thousandth (1/1000) of the equivalent weight of that atom or radical (ion) i.e.,

$$\text{meq. wt.} = \frac{\text{eq. wt. (g)}}{1000}$$

The meq. wt. is often expressed in milligrams (mg), but can also be expressed in grams (g).

*a. Equivalents — examples:**

$$\text{eq. wt. of sodium ion } (Na^{+}), \text{ (atomic wt. 23)} = \frac{23 \text{ g}}{\mathit{1}} = 23 \text{ g}$$

$$\text{eq. wt. of sodium chloride molecule (NaCl), (M.W. 58.5)} = \frac{58.5 \text{ g}}{\mathit{1}} = 58.5 \text{ g}$$

$$\text{eq. wt. of magnesium ion } (Mg^{2+}) \text{ (atomic wt. 24)} = \frac{24 \text{ g}}{\mathit{2}} = 12 \text{ g}$$

To determine the equivalent weight of a compound such as magnesium chloride hexahydrate ($MgCl_2 \cdot 6H_2O$), the molecular weight (M.W.) of the compound in grams (or gram-molecular weight) is divided by the product of the valence of either pertinent ion and the number of times this ion occurs in one molecule of the compound. For example:

* Valence values are in *italics* to facilitate following their role in calculations.

$$MgCl_2 \cdot 6H_2O \rightarrow Mg^{2+} + 2\ Cl^- + 6\ H_2O \text{ (M.W. 203)}$$

$$\text{eq. wt. of } MgCl_2 \cdot 6H_2O = \frac{\text{M.W. (g)}}{\text{valence of specified ions} \times \text{no. of specified ions in 1 mol salt}}$$

$$= \frac{203}{2 \times 1} = 102 \text{ g/eq. of } Mg^{2+}$$

$$\text{and} \quad = \frac{203}{1 \times 2} = 102 \text{ g/eq. of } Cl^-$$

b. Milliequivalents — examples:

$$\text{meq. wt. of sodium } (Na^+) \text{ (atomic wt. 23)} = \frac{23\text{ g}}{1000} = 23 \text{ mg}$$

$$\text{meq. wt. of calcium } (Ca^{2+}) \text{ (atomic wt. 40)} = \frac{20\text{ g}}{1000} = 20 \text{ mg}$$

$$\text{meq. wt. of chloride } (Cl^-) \text{ (atomic wt. 35.5)} = \frac{35.5\text{ g}}{1000} = 35.5 \text{ mg}$$

$$\text{meq. wt. of bicarbonate radical } (HCO_3^-) \text{ (M.W. 61)} = \frac{1 + 12 + (3 \times 16)\text{ g}}{1000} = \frac{61\text{ g}}{1000} = 61 \text{ mg}$$

The weight of a salt, in milligrams, containing 1 meq. of a specified ion

$$= \frac{\text{M.W. salt used (include water of hydration)}}{\text{valence of specified ion} \times \text{no. of specified ions in 1 mol salt}}$$

Examples:

1. Given calcium chloride hexahydrate B.P. ($CaCl_2 \cdot 6H_2O$) (M.W. 219), find the weight (mg) of $CaCl_2 \cdot 6H_2O$ containing 1 meq. of Ca^{2+} and 1 meq. of Cl^-.
   Using the above formula,

$$\text{wt. of salt (mg)} = \frac{\text{M.W. of salt (mg)}}{\text{valence of specified ion} \times \text{no. of specified ions in 1 mol salt}}$$

$$= \frac{219}{2 \times 1} = 109.5 \text{ mg of } CaCl_2 \cdot 6H_2O \text{ contains 1 meq. of } Ca^{2+}$$

$$\text{and} \quad \frac{219}{1 \times 2} = 109.5 \text{ mg of } CaCl_2 \cdot 6H_2O \text{ contains 1 meq. of } Cl^-$$

2. Given calcium chloride dihydrate ($CaCl_2 \cdot 2H_2O$) (M.W. 147), find the weight of $CaCl_2 \cdot 2H_2O$ containing 1 meq. of Ca^{2+} and 1 meq. of Cl^-.

$$\frac{147}{2 \times 1} = 73.5 \text{ mg of } CaCl_2 \cdot 2H_2O \text{ contains 1 meq. of } Ca^{2+}$$

and

$$\frac{147}{1 \times 2} = 73.5 \text{ mg of } CaCl_2 \cdot 2H_2O \text{ contains 1 meq. of } Cl^-$$

Alternatively, the weight of the salt containing 1 meq. of a specified ion may be calculated by dividing the sum of the atomic weights of the constituent atoms, expressed in milligrams (i.e., molecular weight of the salt in milligrams), by the product of the valence of the specified ion times the number of specified ions in 1 mol of the salt. For example, to prepare a solution of calcium chloride using $CaCl_2 \cdot 6H_2O$ that is to contain 1 meq. of Cl^- and 1 meq. of Ca^{2+}, the weight of salt needed

and

$$= \frac{40 + (2 \times 35.5) + 6\,(2 \times 1 + 16)\text{ mg}}{1 \times 2}$$

$$= 109.5 \text{ mg to supply 1 meq. of } Cl^-$$

and

$$= \frac{40 + (2 \times 35.5) + 6\,(2 \times 1 + 16)\text{ mg}}{2 \times 1} = 109.5 \text{ mg to supply 1 meq. of } Ca^{2+}$$

The milliequivalents of an ion in a solution can be calculated by dividing the weight of the salt in solution by the milliequivalent weight (expressed in grams) of the specified ion in the salt, i.e.,

$$\text{meq.} = \frac{\text{wt. (g)}}{\text{meq. wt. (g)}}$$

Similarly, the weight of a salt in a given number of milliequivalents can also be calculated using the above equation, i.e., weight of salt (g) = milliequivalents × milliequivalent wt. (g). The following calculations serve as examples:

1. The number of milliequivalents of calcium and of chloride in 7.35 g of calcium chloride dihydrate ($CaCl_2 \cdot 2H_2O$, M.W. 147) may be calculated in two ways.

Calculation I:

$$\text{meq.} = \frac{\text{wt. (g)}}{\text{meq. wt. (g)}}$$

$$\text{meq. wt.} = \frac{\text{M. W. salt}}{\text{valence of specified ion} \times \text{no. of specified ions in 1 mol salt}}$$

For calcium:

$$\text{the meq. wt.} = \frac{147}{2 \times 1} = 73.5 \text{ (mg)} = 0.0735 \text{ g}$$

$$\therefore \text{meq. } Ca^{2+} = \frac{7.35 \text{ (g)}}{0.0735 \text{ (g)}} = 100 \text{ meq.}$$

For chloride:

$$\text{the meq. wt.} = \frac{147}{1 \times 2} = 73.5 \text{ (mg)} = 0.0735 \text{ g}$$

$$\therefore \text{meq. } Cl^- = \frac{7.35 \text{ (g)}}{0.0735 \text{ (g)}} = 100 \text{ meq.}$$

Thus there are 100 meq. of Ca^{2+} and of Cl^- in 7.35 g of $CaCl_2 \cdot 2H_2O$.

Calculation II (alternative):
Atomic wt. of calcium = 40, of chloride = 35.5

$\therefore$ quantity of calcium in the 7.35 g of $CaCl_2 \cdot 2H_2O$

$$= \frac{40}{147} \times 7.35 = 2.0 \text{ g}$$

$$\therefore 1 \text{ meq. of } Ca^{2+} = \frac{\text{atomic wt. (mg)}}{\text{valence}} = \frac{40}{2} = 20 \text{ mg}$$

$$\therefore \text{meq. of } Ca^{2+} \text{ in } 7.35 \text{ g } CaCl_2 \cdot H_2O = \frac{2\text{g} \times 1000}{20 \text{ mg}} = \frac{2000 \text{ mg}}{20 \text{ mg}} = 100 \text{ meq.}$$

$$\text{Similarly, chloride content} = \frac{2 \times 35.5}{147} \times 7.35 = 3.55 \text{ g},$$

$$\therefore 1 \text{ meq. of } Cl^- = \frac{\text{atomic wt. (mg)}}{\text{valence}} = \frac{35.5 \text{ mg}}{1} = 35.5 \text{ mg}$$

$$\therefore \text{meq. of } Cl^- \text{ in } 7.35 \text{ g } CaCl_2 \cdot 2H_2O = \frac{3.55 \text{ g} \times 1000}{35.5 \text{ mg}} = \frac{3550 \text{ mg}}{35.5 \text{ mg}} = 100 \text{ meq.}$$

2. How many grams of sodium chloride are present in a solution containing 10 meq. of sodium?

Calculation:

$$\text{(a) meq. wt. (g)} = \frac{\text{M. W. salt}}{\text{valence of specified ion} \times \text{no. of specified ions in 1 mol salt}}$$

$$\therefore \text{meq. wt. } Na^+ = \frac{58.5}{1 \times 1} = 58.5 \text{ mg} = 0.0585 \text{ g}$$

$$\text{(b) meq.} = \frac{\text{wt. salt (g)}}{\text{meq. wt. (g)}}$$

$$10 = \frac{\text{wt. salt}}{0.0585} \text{ (g)}$$

$$\text{wt. salt} = 10 \times 0.0585$$

$$= 0.585 \text{ g}$$

$\therefore$ 0.585 g sodium chloride contains 10 meq. sodium

Milliequivalents per liter can also be expressed as weight per liter. The two following formulas show the relationship between concentrations expressed in milliequivalents per liter and as weight per liter:

$$\text{(c) meq. per liter} = \frac{\text{wt. in mg/}\ell \times \text{valence}}{\text{ionic wt. or formula wt. or M.W.}}$$

$$\text{(d) weight (mg/}\ell) = \frac{\text{meq./}\ell \times \text{ionic wt. or formula wt. or M.W.}}{\text{valence}}$$

Examples:

1. Calculate the milliequivalents of sodium and chloride ions in 1000 mℓ of 0.45% (0.45 g/100 mℓ, 450 mg/100 mℓ, 4500 mg/ℓ) sodium chloride injection.

Calculation I:

$$\text{meq. per liter} = \frac{\text{wt. (mg) per liter} \times \text{valence}}{\text{M.W. NaCl}}$$

$$\therefore \text{the meq./}\ell = \frac{4500 \times 1}{58.5}$$

$$= 76.9$$

Each liter contains 76.9 meq. of sodium and 76.9 meq. of chloride.

Calculation II:
100 mℓ of 0.45% NaCl contains 0.45 g NaCl (M.W. 58.5).

$$\text{1 eq. wt.} = \frac{\text{M.W. salt (g)}}{\text{valence of specified ion} \times \text{no. of specified ions in 1 mol salt}}$$

$$\therefore \text{1 eq. wt.} = \frac{58.5\ \text{g}}{1 \times 1} = 58.5\ \text{g and 1 meq. wt.} = \frac{58.5\ \text{g}}{1000} = 0.0585\ \text{g}$$

As before, the number of meq.

$$= \frac{\text{wt. salt (g)}}{\text{meq. wt. (g)}}$$

$$= \frac{0.45\ \text{(g)}}{0.0585\ \text{(g)}}$$

$$= 7.69\ \text{meq. } Na^+ \text{ and } Cl^-/100\ \text{m}\ell$$

$$= 76.9\ \text{meq. } Na^+ \text{ and } Cl^-/1000\ \text{m}\ell$$

2. How much potassium chloride salt will be needed to prepare 1000 mℓ of solution containing 40 meq. of each ion. Molecular weight of potassium chloride (KCl) = 74.6.

Calculation I:

$$\text{Weight in mg/}\ell = \frac{\text{meq. per liter} \times \text{M.W.}}{\text{valence}}$$

$$= \frac{40 \times 74.6}{1}$$

$$= 2984.0\ \text{mg or } 2.984\ \text{g}$$

Therefore 2.984 g of potassium chloride will be needed.

Calculation II:
M.W. of KCl = 74.6

$$\text{1 meq. of KCl} = \frac{74.6}{1 \times 1} = 74.6\ \text{mg/meq. } K^+ \text{ and } Cl^-$$

Table 1 following gives the milliequivalent weight of various ions and compounds (in g) and Table 2 the relationship between milliequivalents and millimoles of some commonly used electrolyte salts.

III. MOLES (mol) AND MILLIMOLES (mmol)

Concentrations of electrolyte and other chemical substances can be expressed in terms other than equivalents or milliequivalents. They can be expressed in moles or millimoles.

A mole (mol)	= one gram-atomic weight or gram-molecular weight of a substance
A millimole (mmol)	= 1/1000 of a mole

Examples:

Sodium chloride has a gram-molecular weight of 58.5 g; thus 1 mol = 58.5 g of sodium chloride. A 1 *M* (mol/*ℓ*) solution of sodium chloride contains 58.5 g of salt per liter of solution and 1 millimolar (mmol/*ℓ*) solution contains 58.5 mg of salt per liter of solution. Molar concentrations of solutions always refer to the number of moles per liter of solution.

To convert the given weight of a salt into moles or millimoles, one divides the weight in grams or milligrams by the molecular weight of the salt.

Examples:

$$\text{5 g of sodium chloride} = \frac{5}{58.5} = 0.08547 \text{ mol}$$

$$= 85.5 \text{ mmol NaCl}$$

IV. MILLIMOLES AND MILLIEQUIVALENTS

There is a relationship between millimoles and milliequivalents (Table 2).

Example 1:

$$\text{Sodium chloride (M.W. 58.5)}: \underset{\text{(salt)}}{NaCl} \rightleftharpoons \underset{\text{(ions in solution)}}{Na^+ + Cl^-}$$

$$\text{1 mmol of NaCl} = \text{M.W. in mg} = 58.5 \text{ mg}$$

$$\text{but 1 meq.} = \frac{\text{M.W. (mg)}}{\text{valence of an ion} \times \text{no. of ions}}$$

$$\text{and 1 meq. } Na^+ \text{ is contained in } \frac{58.5}{1 \times 1} = 58.5 \text{ mg NaCl}$$

$$\text{and 1 meq. } Cl^- \text{ is contained in } \frac{58.5}{1 \times 1} = 58.5 \text{ mg NaCl}$$

$\because$ 1 mmol Na^+ is contained in 1 M.W. NaCl in (58.5) mg, and

$\because$ 1 meq. Na^+ is also contained in 58.5 mg NaCl

$$\therefore \text{1 mmol } Na^+ = 58.5 \text{ mg of NaCl}$$

$$= 1 \times 58.5 \text{ mg}$$

$$= 1 \times \text{meq. wt.}$$

Table 1
MILLIEQUIVALENT WEIGHTS

Ion or compound	meq. weight
Sodium	0.023
Potassium	0.039
Calcium	0.020
Magnesium	0.012
Ammonium	0.018
Chloride	0.0355
Lactate	0.089
Citrate	0.063
Acetate	0.059
Bicarbonate	0.061
Sodium chloride	0.0585
Potassium chloride	0.0746
Calcium chloride, anhydrous	0.0555
Calcium chloride, dihydrate	0.0735
Magnesium chloride hexahydrate	0.102
Ammonium chloride	0.0535
Magnesium sulfate heptahydrate	0.123
Sodium lactate	0.112
Sodium citrate dihydrate	0.098
Sodium acetate trihydrate	0.136
Sodium bicarbonate	0.084

Reproduced from Turco, S. and King, R. E., *Sterile Dosage Forms,* 2nd ed., Lea & Febiger, Philadelphia, 1979, 260. With permission.

Example 2:

Calcium chloride anhydrous (M.W. 111) : $\underset{\text{(salt)}}{CaCl_2} \rightleftharpoons \underset{\text{(ions in solution)}}{Ca^{2+} + 2\ Cl^-}$

1 mmol of $CaCl_2$ = M.W. (mg) = 111 mg

$$\text{but 1 meq. of } Ca^{2+} = \frac{\text{M.W. (mg)}}{\text{valence of ion} \times \text{no. of ions}}$$

$$= \frac{111}{2 \times 1} = 55.5 \text{ mg of } CaCl_2$$

$$\text{and 1 meq. of } Cl^- \text{ is contained in } \frac{111}{1 \times 2}$$

$$= 55.5 \text{ mg of } CaCl_2$$

$\because$ 1 mmol of Ca^{2+} is contained in 1 M.W. $CaCl_2$ (111 mg), and

$\because$ 1 meq. of Ca^{2+} is contained in 55.5 mg $CaCl_2$

$\therefore$ 1 mmol of Ca^{2+} = 111 mg of $CaCl_2$ = 2 × 55.5 mg

= 2 × meq. wt.

The number of millimoles (mmol) of an ion can be seen to be equal to the number of milliequivalents of that ion divided by the valence of that ion, i.e.,

Table 2
RELATIONSHIP BETWEEN meq. AND mmol OF SOME COMMONLY USED ELECTROLYTE SALTS

Electrolyte salt	Formula	Mol. wt. of salt (mg)	meq. wt. of salt (mg)	Type of ion	No. of meq. of ion/meq. wt. of salt	No. of mmol of ion/meq. wt. of salt
		SODIUM (Na^+)				
Sodium phosphate monobasic monohydrate	$NaH_2PO_4 \cdot H_2O$	138.0	138.0	Na^+	1	1
				HPO_4^{2-}	1	1
Sodium phosphate dibasic	Na_2HPO_4	141.98	70.99	Na^+	1	1
				HPO_4^{2-}	1	0.5
Sodium bicarbonate	$NaHCO_3$	84.0	84.0	Na^+	1	1
				NCO_3	1	1
Sodium chloride	NaCl	58.5	58.5	Na^+	1	1
				Cl^-	1	1
Sodium lactate	$NaC_3H_5O_3$	112.07	112.07	Na^+	1	1
				$C_3H_5O_3^-$	1	1
Sodium acetate trihydrate	$NaC_2H_3O_2 \cdot 3H_2O$	136.0	136.0	Na^+	1	1
				$C_2H_3O_2^-$	1	1
		POTASSIUM (K^+)				
Potassium chloride	KCl	74.56	74.56	K^+	1	1
				Cl^-	1	1
Potassium phosphate monobasic	KH_2PO_4	136.09	136. 09	K^+	1	1
				$H_2PO_4^-$	1	1
Potassium phosphate dibasic	K_2HPO_4	174.18	87.09	K^+	1	1
				HPO_4^{2-}	1	0.5
		MAGNESIUM (Mg^{2+})				
Magnesium sulfate heptahydrate	$MgSO_4 \cdot 7H_2O$	246.5	123	Mg^{2+}	1	0.5
				SO_4^{2-}	1	0.5
Magnesium chloride	$MgCl_2 \cdot 6H_2O$	203.3	101.7	Mg^{2+}	1	0.5
				Cl^-	1	1
		CALCIUM (Ca^{2+})				
Calcium chloride dihydrate	$CaCl_2 \cdot 2H_2O$	147	73.5	Ca^{2+}	1	0.5
				Cl^-	1	1
Calcium gluconate monohydrate	$C_{12}H_{22}CaO_{14} \cdot H_2O$	448.4	224.2	Ca^{2+}	1	0.5
				$C_{12}H_{22}O_{14}^{2-}$	1	0.5

$$\text{No. mmol} = \frac{\text{no. meq. of an ion}}{\text{valence of that ion}}$$

or

$$\text{no. of meq.} = \text{no. of mmol} \times \text{valence of given ion}$$

V. CALCULATION OF PHOSPHATE

Phosphate, unlike most other electrolytes, exists in both monovalent ($H_2PO_4^-$) and divalent (HPO_4^{2-}) forms, the ratio of the two depending on the pH of the solution. Because of this variation in the valence of phosphate it is better to express its concentration in terms of elemental phosphorus (mg) or mmol of phosphate as in the following example:

Consider Potassium Phosphate Injection 3 meq. of K^+/mℓ (TGH).

K_2HPO_4 (M.W. 174); ∴ 1 mmol = 174 mg (dibasic form)
KH_2PO_4 (M.W. 136); ∴ 1 mmol = 136 mg (monobasic form)
The given concentration of salt/mℓ is
K_2HPO_4 = 157 mg (dibasic)
KH_2PO_4 = 163.5 mg (monobasic)

Therefore

$$\text{mmol/m}\ell\ K_2HPO_4 = \frac{157}{174} = 0.902 \text{ mmol (dibasic)}$$

$$KH_2PO_4 = \frac{163.5}{136} = 1.202 \text{ (monobasic)}$$

From the above we have

$$\begin{aligned} & 0.902 \text{ mmol of } HPO_4^{2-} \\ + \ & \underline{1.202} \text{ mmol of } H_2PO_4^- \\ \text{Total phosphate} = \ & 2.104 \text{ mmol/m}\ell \end{aligned}$$

Similarly we have

$$\begin{aligned} 2 \times 0.902 &= 1.804 \text{ mmol of } K^+ \text{ from } K_2HPO_4 \\ + 1 \times 1.202 &= \underline{1.202} \text{ mmol of } K^+ \text{ from } KH_2PO_4 \\ \text{Total potassium} &= 3.006 \text{ mmol/m}\ell \end{aligned}$$

$$\because \text{no. of meq.} = \text{no. of mmol} \times \text{valence of specified ion}$$

$$\therefore 3.006 \text{ mmol/m}\ell \text{ of } K^+ = 3.006 \times 1 = 3.006 \text{ meq. of } K^+$$

Since 1 mmol of phosphate radical (ion) provides 1 mmol of phosphorus and since 1 mmol of phosphorus equals 31 mg, then 2.104 mmol of phosphate = 2.104 mmol × 31 = 65.2 mg/mℓ of elemental phosphorus.

VI. TOTAL PARENTERAL NUTRITION CALCULATIONS

Prepare 2 ℓ of TPN solution containing 4.25% amino acids and 25% dextrose with the following electrolytes per liter:

Sodium	60 meq
Potassium	30 meq
Calcium	9 meq
Magnesium	10 meq
Phosphorus	250 meg
Chloride	94 meq
Acetate	37 meq

You are provided with:

- Travasol E.F. 10% Injection, 1000 mℓ (Baxter) E.F. = largely electrolyte-free, but see No. 6 below.
- Dextrose 50% Injection 500 mℓ (Abbott).
- Sodium Chloride 23.4% (4 meq./mℓ) Injection, 250 mℓ (T.G.H.).
- Calcium Gluconate 10% Injection, 10 mℓ (TGH).
- Magnesium Sulfate 50% Injection 10 mℓ (Sterilab.).
- Potassium Phosphate (2.1 mmol/mℓ of phosphate and 3 meq./mmol of K^+/mℓ) Injection, 250 mℓ (TGH).
- Potassium Chloride (4 meq./mℓ) Injection, 250 mℓ (TGH).

Calculate

1. Amount of amino acids in each liter of TPN solution. Since 100 mℓ of solution is to contain 4.25 g amino acids,
   ∴ *1000 mℓ will contain 4.25 × 10 = 42.5 g amino acids*
2. How much Travasol® 10% E.F. will be needed to provide that amount of amino acids per liter? (Travasol® 10% E.F. contains 10 g of amino acids per 100 mℓ of solution.)
   Since 10 g amino acids are in 100 mℓ of Travasol® 10% E.F., then 42.5 g will be in
   100/10 × 42.5 = 425 mℓ of Travasol® 10% E.F.
3. Calculate the amount of nitrogen in 1ℓ of solution if we know that Travasol® *10% contains 16.8g of nitrogen per 100 g of amino acids.*
   Since 100 g of amino acids contain 16.8 g of nitrogen, 42.5 g of amino acids will contain 16.8/100 × 42.5 = 7.14g of nitrogen.
4. What is the approximate protein equivalent (e.g., the equivalent of protein per gram of nitrogen) in 1 ℓ of solution?
   Since the empirical formula (a reasonable approximation for many proteins) states that 1 g of nitrogen = 6.25 g protein
   then 7.14 g nitrogen × 6.25 = 44.6 g protein.
5. What is the ratio of nitrogen (g) to nonprotein energy (kcal) if the patient is receiving this solution?
   The dextrose solution is 25% or 25 g/100 mℓ or 250 g/1000 mℓ. Thus the amount of dextrose per liter of solution is 250 g, and per 2 ℓ = 500 g. Given that 1 g of dextrose provides 3.4 kilocalories (kcal) (since the dextrose is in the form of the monohydrate and so its weight includes some water of hydration which has no caloric value), then 500 g × 3.4 kcal = 1700 kcal.
   The amount of nitrogen in 2 ℓ of solution is 7.14 g × 2 = 14.28 = 14.3 g.

$$\therefore \text{ratio of } \frac{\text{nitrogen(g)}}{\text{nonprotein (kcal)}} = \frac{14.3}{1700} = \frac{1}{119} \text{ or } 1:119$$

6. Calculate the amount of each electrolyte that has to be added to each liter of solution.
   Consider the electrolyte concentration derived from Travasol® E.F. 10% knowing that each liter of this solution contains Na^+ 3 meq., Cl^-40 meq., and acetate 87 meq.
   The 425 mℓ of Travasol® 10% E.F. calculated to be needed in No. 2 (above) will contain:
   Na^+, 0.425ℓ × 3 = 1.3 meq
   Cl^-, 0.425ℓ × 40 = 17 meq
   Acetate, 0.425ℓ × 87 = 37 meq

Since Travasol® 10% E.F. contains no other electrolytes the rest will have to be added to the admixture to provide the required amounts as stated in the prescription.

Electrolytes required:
Sodium: 60 − 1.3 = 58.7 meq

Calculation I:
23.4% NaCl contains 23.4 g of NaCl/100 m*l* of solution.

$$\text{meq. wt. of NaCl} = \frac{\text{M.W. (mg)}}{\text{valence} \times \text{no. of ions}}$$

$$= \frac{58.5}{1 \times 1}$$

$$= 58.5 \text{ mg} = 0.0585 \text{ g}$$

$$\text{and no. of meq.} = \frac{\text{wt. of salt (g)}}{\text{meq. wt. (g)}}$$

$$= \frac{23.4}{0.0585}$$

$$= 400$$

Therefore there are 400 meq./100 m*l* of NaCl solution. Since we need 58.7 meq. therefore we require 58.7/400 = 14.67 = 14.7 m*l*

Calculation II:

$$\frac{58.7 \text{ meq.}}{4 \text{ meq./m}\ell} = 14.7 \text{ m}\ell \text{ of NaCl of 4 meq./m}\ell$$

CHLORIDE
94 − 17 = 77 meq.
= sufficient from NaCl (59 meq.) + KCl (18 meq.)

ACETATE
37 − 0 = 37 meq. = enough from Travasol®

CALCIUM
9 − 0 = 9 meq, obtained as follows:
Calcium gluconate is $C_{12}H_{22}CaO_{14} \cdot H_2O$ (M.W. 448.4) meq. wt. = molecular weight (mg)/valence × no. of specified ions

$$1 \text{ meq. of } Ca^{2+} = \frac{448.4}{2 \times 1} = 224 \text{ mg}$$

9 meq. of Ca^{2+} = 9 × 224 = 2016 mg.
Calcium gluconate 10% (10 g/100 m*l*) contains 100 mg calcium gluconate per m*l* of solution. The 9 meq. required is contained in 2016 mg of calcium gluconate and this is contained in 2016 mg/100 mg = 20.2 m*l*, or approximately 20 m*l* of calcium gluconate 10% solution.

POTASSIUM
30 − 0 = 30 meq. obtained as follows:

4.6 mℓ of KCl × 4 meq./mℓ	= 18.4 meq. K^+
+ 3.8 mℓ of potassium phosphate × 3.0 meq. K^+/mℓ	= 11.4 meq K^+
Total:	29.8 meq. K^+

MAGNESIUM

10 − 0 = 10 meq. obtained as follows:

$MgSO_4 \cdot 7H_2O$ (M.W. 246.5)

Magnesium sulfate 50% injection (50 g/100 mℓ) contains 500 mg/mℓ

PHOSPHORUS

250 mg − 0 = 250 mg

Since each millimole of phosphate radical in the Potassium Phosphate Injection contains 1 mmol of phosphorus and since 1 mmol of phosphorus is 31 mg, then 2.1 mmol of phosphate/mℓ = 2.1 × 31 = 65.1 mg/mℓ of elemental phosphorus (i.e., each m ℓ of Potassium Phosphate Injection contains 65.1 mg of elemental phosphorus). Therefore 250 mg of phosphorus are contained in 250/65.1 = 3.8 mℓ of Potassium Phosphate Injection 2.1 mmol of phosphorus/mℓ.

On the other hand, if we were asked to add 12 meq. of phosphate to the solution instead of receiving a request for phosphate as elemental phosphorus (250 mg), we could calculate approximately how much potassium phosphate to add (in millimoles) by using the following equation and assuming that all the phosphate is in the divalent form. This, of course, is only an approximation.

$$\text{no. of mmol} = \frac{\text{meq.}}{\text{valence of ion}}$$

$$= \frac{12 \text{ meq.}}{2}$$

$$= 6 \text{ mmols of } PO_4^{2-}$$

Therefore the volume of Potassium Phosphate Injection 2.1 mmol/mℓ that is needed = 6 mmol/2.1 mmol = 2.8 mℓ.

7. How much dextrose is needed to provide a 25% dextrose concentration in the final solution?

   Quantity of dextrose (as dextrose monohydrate) per liter of solution is 0.25 g × 1000 = 250 g. From the 50% dextrose concentrate it is seen that 50 g of dextrose are in 100 mℓ of solution, ∴ 250 g will be in 100/50 × 250 = 500 mℓ of 50% dextrose concentration.

8. Assuming that the concentration of dextrose in the final solution will be constant at 25%, what would be the volume of Travasol® 10% in the final solution of 1ℓ that is necessary to supply 200 kcal/g of nitrogen?

   The manufacturer states that Travasol® 10% contains 16.8 g of nitrogen per liter of solution.

   Since 1ℓ of the TPN admixture contains 0.025 × 1000 = 250 g of dextrose monohydrate and since 1 g of dextrose monohydrate provides 3.4 kcal, then 250 g will provide 250 × 3.4 = 850 nonprotein kcal.

   To achieve the desired ratio of 200 nonprotein kcal/g of nitrogen, 850 kcal will require 850/200 = 4.25 g of nitrogen, and given that 100 mℓ of Travasol® 10% contains 1.68 g of nitrogen:

∴ volume of Travasol® 10% required for 4.25 g of nitrogen = 100/1.68 × 4.25 = 253 mℓ,

∴ 253 mℓ of Travasol® 10% contains 25.3 g of amino acids,

∴ the percent of Travasol® in the final TPN admixture will be 25.3/1000 × 100 = 2.53%, and

∴ a solution containing 2.53% Travasol® and 25% dextrose will provide 200 non-protein kcal/g of nitrogen.

Chapter 22

COMMON TERMINOLOGY

George Tsallas

1. **Total parenteral nutrition (TPN)** is a special procedure using either central or peripheral veins for the intravenous administration of sterile solutions of amino acids, carbohydrates, minerals, vitamins, and fat, to meet (total) nutritional requirements for normal growth and metabolism. Synonymous terms include intravenous hyperalimentation, parenteral nutrition, or parenteral alimentation.
2. **Central parenteral nutrition (central intravenous feeding)** is the intravenous administration of sterile hypertonic solutions of amino acids, carbohydrates, minerals, vitamins, and fat through a central catheter, the tip of which is advanced into the vena cava. By reason of the substantial dilution at this site, solutions of high osmolality (1500 to 1700 mOsm/kg) may be infused.
3. **Peripheral parenteral nutrition (peripheral intravenous feeding)** is the intravenous administration of less hypertonic solutions of amino acids, carbohydrates, minerals, and vitamins (approximately 300 to 700 mOsm/kg) given concomitantly with an isotonic sterile fat emulsion through a peripheral vein.
4. **An intravenous (i.v.) admixture** is the combination of several different sterile components into a single sterile large volume unit for intravenous infusion.
5. **A large volume parenteral (L.V.P.)** is a sterile solution, excluding blood or blood fractions, of 100 mℓ or more, intended for intravenous injection and used in the diagnosis or treatment of disease or modification of physiological functions.
6. **A protein hydrolysate** is a sterile solution of hydrolyzed natural protein such as casein or fibrin that is prepared by hydrolysis of the protein using either enzymes or hydrochloric acid. It contains essential, semiessential and nonessential amino acids as well as di- and tripeptides.
7. **A (defined) crystalline amino acid solution** is a manufactured sterile amino acid solution of pure, synthetically derived L-amino acids in which known concentrations of essential, semiessential and nonessential amino acids are present.
8. **A sterile fat emulsion** is a sterile isotonic injection of an emulsion manufactured from the vegetable oil of soybean or safflower, intended for intravenous administration and supplying both essential fatty acids and calories.
9. **Contaminants** are substances introduced accidentally and may go unrecognized. They may be of organic or inorganic nature, including trace chemicals, microorganisms, pyrogens, and particulate matter that are introduced into a sterile preparation during the manufacture, reconstitution, storage, and administration of the product.
10. **Sterilization** is any process by means of which all forms of microbial life (bacteria, spores, fungi, and viruses) contained in liquids, on instruments and utensils, or within various substances, are completely destroyed.
11. **Pyrogens** are toxic metabolic byproducts from bacteria and have a complex lipopolysaccharide nature. When infused into the bloodstream they cause fever, sweating, rigors, and other undesirable clinical symptoms.
12. **Inline-filters** are sterile membrane filter units with a pore size varying from 0.45 to 0.22 μm attached inline along the infusion tube between the container of the

amino acid-dextrose-electrolyte mixture and the patient, to retain microbiological contaminants and particulate matter that may have been introduced into the solution either during the preparation of the admixture or administration of the solution.

13. **Osmotic pressure** is the force created by dissolved particles across a semipermeable membrane, which results in the passage of water from a region of greater concentration to one of lesser concentration.
14. **Osmosis** is the movement of solvent molecules through a semipermeable membrane from a solution of lower solute concentration to one of higher solute concentration.
15. **Osmolarity** is the total osmotic force of all solutes dissolved per liter of solution. It is expressed as the number of osmoles or milliosmoles per liter of solution.
16. **Osmolality** is an expression of the osmotic force expressed as the number of osmoles or milliosmoles per kilogram of water (this being the mass of the solvent).
17. **Osmoles and milliosmoles** are units expressing the osmotic pressure of a solution, for which concentration is in terms of molality or molarity of the solutions (e.g., moles per kilogram of solvent or moles per liter of solution, respectively); thus osmoles or milliosmoles per kilogram of solvent (water) and osmoles or milliosmoles per liter of solution for dilute solutions.
18. **Tonicity** refers to the effective osmotic pressure exerted by a solution due to the solutes or dissolves solids present, relative to that of plasma.
19. **An isotonic solution** is one having the same osmotic pressure or the same concentration of dissolved solids as that of plasma. Cells in contact with a solution of this tonicity will not change physically.
20. **Hypotonic** refers to a solution having a lower osmotic pressure or a lower concentration of dissolved solids than that of plasma. Cells in contact with hypotonic solutions may swell and burst.
21. **Hypertonic** refers to a solution having a higher osmotic pressure or higher concentration of dissolved solids than that of plasma. Cells in contact with a hypertonic solution may shrink and lose some of their fluid content.
22. **pH** is the inverse of the logarithm of the hydrogen ion concentration and indicates the degree of acidity or alkalinity of a solution. The pH values range from 0 to 14, with values lower than 7 indicating acidity and values greater than 7 indicating alkalinity.
23. **Electrolytes** are salts of either acids or bases which when dissolved in water will form cations (positively charged ions) and anions (negatively charged ions) capable of conducting an electrical current. Electrolytes are important in maintaining a normal acid-base balance of the body, and osmolarity, and are necessary for membrane function and nutrient metabolism.
    Example:

    $$\underset{}{NaCl} \rightleftharpoons \underset{\text{(cation)}}{Na^+} + \underset{\text{(anion)}}{Cl^-}$$

    Salt dissolved in water is dissociated into sodium ions (Na^+) and chloride ions (Cl^-).
24. **Equivalents and milliequivalents** are the mass of an element expressed as a ratio of the molecular or atomic weight divided by the valency. The importance of expressing chemicals in this unit is that elements react in combinations which are multiples of their equivalent weights.
25. **Moles and millimoles** are the mass of an element expressed in terms of their molecular weight taken as grams or milligrams.
26. **Instability** consists of changes in the physical, chemical, or therapeutic character-

istics of an intravenous admixture which result in a loss of 10% or more of one of the constituents claimed on its label and renders the product unsuitable for administration.

27. **Incompatibility** is an undesirable physical, chemical, or functional effect caused by mixing two substances or administrating them together. This may be

- Therapeutic: when two or more drugs are administered concurrently, resulting in an undesirable antagonistic or synergistic pharmacologic action.
- Physical: when a drug in solution causes a change in the appearance of the solution such as in color, in clarity, or in the evolution of gas.
- Chemical: when a drug becomes ineffective due to a chemical reaction, visible or invisible, with the admixture.

28. **Compatibility** is the admixing of two or more drugs without causing undesirable physical, chemical, or therapeutic effects.
29. **The calorie** is a unit of thermal energy (unit of heat) equivalent to 4.184 joules (J), required to raise the temperature of one gram of water by one degree Celsius at a standard pressure of one atmosphere. When applied to a study in metabolism it is referred to as the large calorie or kilocalorie (kcal), and is equivalent to the 4.184 kilojoules (KJ) that are required to raise the temperature of one kilogram of water by one degree Celsius.
30. **Piggy back** is the administration of a second solution intermittently or by continuous infusion through a Y tube or Y-injection site attached to the primary administration set.
31. **Aseptic technique** is the use of procedures in the preparation and administration of parenteral products which minimize or prevent the introduction of microorganisms. This process demands that bare hands or unsterile objects do not touch sterile areas. It also includes avoidance of contamination by airborn bacteria through the use of masks and laminar air-flow hoods.

READING LIST

Bergman, D. H., Incompatibilities in large volume parenterals, *Drug Intell. Clin. Pharm.*, 11, 345, 1977.

The Committee for the Standardization of Intravenous Terminology and Parenteral Admixture Guidelines, Canadian Society Hospital Pharmacists, Toronto, 1980.

Heller, W. M., *U.S. Pharmacopeia*, 19th rev., U.S. Pharmacopeial Convention, Inc., Rockville, Md., 1975, 582.

Paximos, J. and Samuels, T. M., Combined volume control set — piggy back system for intermittent intravenous, *Ther. Am. J. Hosp. Pharm.*, 32, 892, 1975.

National Coordinating Committee on Large Volume Parenterals, Recommendations for the labelling of large volume parenterals, *Am. J. Hosp. Pharm.*, 35, 49, 1978.

Stanners, J. E., Guidelines for the Intermittent Intravenous Administration of Medications, Pharmacy Residency project, Unviersity of Western Ontario, London 1976.

Trissel, L., A., *Handbook on Injectable Drugs*, American Society of Hospital Pharmacists, Washington, D.C., 1977.

Turco, S. K., Intravenous piggy bank — a new dimension in unit dose, *J. Parenteral Drug Assoc.*, 32, 50, 1978.

Turco, S. and King, R. E., *Sterile Dosage Forms, Their Preparation and Clinical Application*, 2nd ed., Lea & Febiger, Philadelphia, 1979.

INDEX

A

B

D

E

F

G

H

I

N

O

P

Q

R

S

T

U

V

W

Z